IFCoLog Journal of Logics and their Applications

Volume 3, Number 5

December 2016

Disclaimer
Statements of fact and opinion in the articles in IfCoLog Journal of Logics and their Applications are those of the respective authors and contributors and not of the IfCoLog Journal of Logics and their Applications or of College Publications. Neither College Publications nor the IfCoLog Journal of Logics and their Applications make any representation, express or implied, in respect of the accuracy of the material in this journal and cannot accept any legal responsibility or liability for any errors or omissions that may be made. The reader should make his/her own evaluation as to the appropriateness or otherwise of any experimental technique described.

© Individual authors and College Publications 2016
All rights reserved.

ISBN 978-1-84890-224-4
ISSN (E) 2055-3714
ISSN (P) 2055-3706

College Publications
Scientific Director: Dov Gabbay
Managing Director: Jane Spurr

http://www.collegepublications.co.uk

Printed by Lightning Source, Milton Keynes, UK

All rights reserved. No part of this publication may be reproduced, stored in a retrieval system or transmitted in any form, or by any means, electronic, mechanical, photocopying, recording or otherwise without prior permission, in writing, from the publisher.

EDITORIAL BOARD

Editors-in-Chief
Dov M. Gabbay and Jörg Siekmann

Marcello D'Agostino	Melvin Fitting	Henri Prade
Natasha Alechina	Michael Gabbay	David Pym
Sandra Alves	Murdoch Gabbay	Ruy de Queiroz
Arnon Avron	Thomas F. Gordon	Ram Ramanujam
Jan Broersen	Wesley H. Holliday	Chrtian Retoré
Martin Caminada	Sara Kalvala	Ulrike Sattler
Balder ten Cate	Shalom Lappin	Jörg Siekmann
Agata Ciabttoni	Beishui Liao	Jane Spurr
Robin Cooper	David Makinson	Kaile Su
Luis Farinas del Cerro	George Metcalfe	Leon van der Torre
Esther David	Claudia Nalon	Yde Venema
Didier Dubois	Valeria de Paiva	Rineke Verbrugge
PM Dung	Jeff Paris	Heinrich Wansing
Amy Felty	David Pearce	Jef Wijsen
David Fernandez Duque	Brigitte Pientka	John Woods
Jan van Eijck	Elaine Pimentel	Michael Wooldridge

Scope and Submissions

This journal considers submission in all areas of pure and applied logic, including:

- pure logical systems
- proof theory
- constructive logic
- categorical logic
- modal and temporal logic
- model theory
- recursion theory
- type theory
- nominal theory
- nonclassical logics
- nonmonotonic logic
- numerical and uncertainty reasoning
- logic and AI
- foundations of logic programming
- belief revision
- systems of knowledge and belief
- logics and semantics of programming
- specification and verification
- agent theory
- databases
- dynamic logic
- quantum logic
- algebraic logic
- logic and cognition
- probabilistic logic
- logic and networks
- neuro-logical systems
- complexity
- argumentation theory
- logic and computation
- logic and language
- logic engineering
- knowledge-based systems
- automated reasoning
- knowledge representation
- logic in hardware and VLSI
- natural language
- concurrent computation
- planning

This journal will also consider papers on the application of logic in other subject areas: philosophy, cognitive science, physics etc. provided they have some formal content.

Submissions should be sent to Jane Spurr (jane.spurr@kcl.ac.uk) as a pdf file, preferably compiled in LaTeX using the IFCoLog class file.

CONTENTS

ARTICLES

String Unification is Essentially Infinitary 755
Michael Hoche, Jörg Siekmann and Peter Szabo

Quantum States and Disjunctive Attacks in Talmudic Logic 789
Michael Abraham, Israel Belfer, Uri Schild and Dov Gabbay

More Modal Semantics without Possible Worlds 815
Hitoshi Omori and Daniel Skurt

Three Grades of Agnostic Involvement 847
Gillman Payette

Intuitionistic Modal Logic made Explicit 877
Michel Marti and Thomas Studer

The Indian Schema Analogy Principles 903
J. B. Paris and A. Vencovská

String Unification is Essentially Infinitary

Michael Hoche
Airbus Defense and Space, Claude-Dornier-Strasse, D-88090 Immenstaad

Jörg Siekmann
Saarland University/DFKI, Stuhlsatzenhausweg, D-66123 Saarbrücken

Peter Szabo
Kurt-Schumacher-Str. 13, D-75180 Pforzheim

Abstract

A *unifier* of two terms s and t is a substitution σ such that $s\sigma = t\sigma$ and for first-order terms there exists a *most general unifier* σ in the sense that any other unifier δ can be composed from σ with some substitution λ, i.e. $\delta = \sigma \circ \lambda$.

For many practical applications it turned out to be useful to generalize this notion to E-unification, where E is an equational theory, $=_E$ is equality under E and σ is an E-unifier if $s\sigma =_E t\sigma$. Depending on the equational theory E, the set of most general unifiers is always a singleton (as above) or it may have more than one unifier, either finitely or infinitely many unifiers and for some theories it may not even exist, in which case we call the theory of type nullary.

String unification (or Löb's problem, Markov's problem, unification of word equations or Makanin's problem as it is often called in the literature) is the E-unification problem, where $E = \{f(x, f(y,z)) = f(f(x,y),z)\}$, i.e. unification under associativity or string unification once we drop the fs and the brackets. It is well known that this problem is infinitary and decidable.

Essential unifiers, as introduced by Hoche and Szabo, generalize the notion of a most general unifier and have a dramatically pleasant effect in the sense that the set of essential unifiers is often much smaller than the set of most general unifiers. Essential unification may even reduce an infinitary theory to

---

We would like to thank our first reviewer at the unification workshop in 2008 and the interesting discussion there and afterwards with several participants of the workshop. All of this led to a complete reformulation of our basic definitions and greatly simplified the proofs and the general presentation, finally leading to our more general framework based on the encompassment order as presented here and in [67]. We also acknowledge the very critical and competent later reviews of this paper. We are also indebted to Artur Jez' substantial contribution to paragraph 3.1, where he pointed to a serious flaw in our first version of this paper (the unitary, finitary result)

an essentially finitary theory. The most dramatic reduction known so far is obtained for idempotent semigroups or bands as they are called in computer science: bands are of type nullary, i.e. there exist two unifiable terms s and t, for which the complete and minimal set of most general unifiers does not exist. This is in stark contrast to essential unification: the set of essential unifiers for bands always exists and is finite.

We show in this paper that string unification in one variable, known to be infinitary, has a finite number of essential unifiers (i.e. is e-finitary), however the early hope for a similar reduction of unification under associativity is not justified: string unification is *essentially infinitary*. We give an enumeration algorithm for essential unifiers.

Keywords: *E*-unification, equational theory, essential unifiers, string unification, word equations, universal algebra, unification algorithms.

1 Introduction

Unification is a well established concept in artificial intelligence, automated theorem proving, the semantic web, in computational linguistics and in universal algebra as well as in theoretical and applied computer science like for example in semantics of programming languages (see [56, 42, 65] for several application areas). Surveys of unification theory can be found in [6, 7, 24, 42, 65]. A survey of the related topic of rewriting systems is presented in [15] and in the "emerging" textbook [41]; a list of open problems can be found in [1]. A standard textbook is by Franz Baader and Tobias Nipkow, *Term Rewriting and All That* [5]. A recent survey on higher order unification is [34].

Unification is a general mechanism to solve equational problems. It is in particular embedded in a plurality of deduction and inference mechanisms and for practical applications it is often crucial to have a finite or at least minimal representation of all the solutions, i.e. a minimal complete set of unifiers from which all other solutions (unifiers) can be derived. For unification problems in the free algebra of terms (also known as syntactic unification), there exists always a single unifier for solvable unification problems from which all other unifiers can be derived by instantiation. This unique (up to renaming) unifier is called the *most general unifier* [57]. However for equational algebras the situation is completely different: the minimal complete set of unifiers is not always finite and it may not even exist, which was conjectured by Gordon Plotkin in his seminal paper in 1972 [55]. Since then unification problems and equational theories have been classified with respect to the cardinality of their minimal complete set of unifiers. These results led to the development of general approaches and algorithms, which can be applied to a whole class of theories [19, 9, 22] and others. This is the topic of *universal unification*, see e.g. [69, 64].

More specifically, an *E*-unification problem $s =^?_E t$ for two given terms s and t over a signature Σ and an equational theory E is the problem to find a minimal and complete set of unifiers $\mu\mathcal{U}\Sigma_E$ for s and t, such that for every $\sigma \in \mu\mathcal{U}\Sigma_E$ we have $s\sigma =_E t\sigma$ (i.e. correctness) and for any unifier δ there is a σ in $\mu\mathcal{U}\Sigma_E$ and some substitution λ, such that $\delta =_E \sigma \circ \lambda$ (i.e. completeness). $\mu\mathcal{U}\Sigma_E$ is also minimal in the sense that for every two unifiers σ,τ in $\mu\mathcal{U}\Sigma_E$ there is no λ with $\sigma =_E \tau \circ \lambda$, i.e. all unifiers in $\mu\mathcal{U}\Sigma_E$ are independent. We say a unification problem is *unitary* if $\mu\mathcal{U}\Sigma_E$ is always a singleton, it is *finitary* if $\mu\mathcal{U}\Sigma_E$ is finite for every s and t and it is *infinitary* if there are terms s and t such that $\mu\mathcal{U}\Sigma_E$ is infinite. Unfortunately there are theories such that two terms are unifiable, but the set $\mu\mathcal{U}\Sigma_E$ does not exist. In this case we call the problem *nullary* or of *type zero*. This classification according to the type naturally leads to a hierarchy of equational theories called the *unification hierarchy*.

It turned out that this well established view of unification theory changes drastically, if we redefine the notion of a most general unifier. Recall that a unifier σ *subsumes* another unifier τ if:

$$\tau =^V_E \sigma \circ \lambda$$

Hence standard unification theory is based on the *subsumption relation*. We generalize this notion and define *an encompassment relation* on substitutions: a substitution σ is encompassed by a substitution τ, if there exist substitutions λ_1 and λ_2 such that

$$\tau =^V_E \lambda_1 \circ \sigma \circ \lambda_2$$

where λ_1 has to have certain properties to be defined in the next paragraph below. The idea is that λ_2 is used to establish the known subsumption relation between τ and σ as in standard unification theory and is composed as usual "from the right" in the tripartition $\lambda_1 \circ \sigma \circ \lambda_2$. The substitution λ_1 allows us also to compose "from the left" and this can drastically reduce the cardinality of the set of most general E-unifiers, which we now call *essential E-unifiers*: an E-unifier τ is an essential E-unifier if there is no E-unifier σ with substitutions λ_1 and λ_2 such that $\tau =^V_E \lambda_1 \circ \sigma \circ \lambda_2$. We say τ *encompasses* σ and the *set of essential E-unifiers*, denoted as $e\mathcal{U}\Sigma_E$, is the set of E-unifiers such that for any unifier τ there is some $\sigma \in e\mathcal{U}\Sigma_E$, such that $\tau =^V_E \lambda_1 \circ \sigma \circ \lambda_2$.

The set of essential unifiers - for nonnullary theories - is in particular always a subset of the set of most general unifiers. So by analogy we say a unification problem is essentially unitary, i.e. it is *e-unitary* (it is *e-finitary*) if the set of essential unifiers is always a singleton (is always finite). A unification problem is *e-infinitary* (*e-nullary*) if there are two terms such that the set of essential unifiers is infinite (does not exist).

These notions were first introduced by Hoche and Szabo in [31] where it was shown that the unification problem for idempotent semigroups (bands) is *e*-finitary. Bands are well known since it was one of the early examples to demonstrate Plotkin's conjecture, that there exist nullary equational theories, which was shown one and a half decades later independently by Gerard Huet [23], Manfred Schmidt-Schauss [59] and Franz Baader [3], see also [4]. Now the unification problem for bands is nullary in the traditional sense but it is *e*-finitary in our sense: this is so far the most drastic reduction of the cardinality of the set of most general unifiers to a set of essential unifiers.

The question is: can similar results be obtained for other theories as well. A natural candidate for this kind of investigation is string unification. Why is that?

In the 1950s A. A. Markov was interested in Hilbert's 10th problem and tried to reduce it to the solvability of word equations in free semigroups [32]: he noted that every word equation over a two constant alphabet can be translated into a set of diophantine equations [51]. Using this translation he hoped to find a proof for the unsolvability of Hilbert's tenth problem by showing that the solvability of word equations is undecidable [52]. This put the problem firmly on the map and others joined in: Lentin and Schützenberger [46], J. I. Hmelevskij [27, 28, 29], V. K. Bulitko [10], A. Lentin [45], V. G. Durnev [21, 20] and many others, see [2] for a survey as well as the volumes edited by several mathematicians under the pseudonym of M. Lothaire on *Algebraic Combinatorics on Words* [48, 49]. The problem was finally solved in the affirmative in the seminal work by G. S. Makanin [50]. An exposition of Makanin's algorithm (with several improvements) is presented inter alia by Klaus Schulz [61, 60] and by Volker Diekert [16]. An algorithm for the computation of a minimal and complete set of unifiers is given in [36] and there is a history of improved algorithms and their complexity bounds, some standard references are e.g. [43], [25] and an algorithm different to Makanin is [54]. Some articles on special cases (strings with one or two variables) are [11, 44, 53] and [37, 38] and a recent generalization with up to date references is [17]. Since then the amount of works and results for this and related problems has exploded even more.[1]

Apart from its theoretical and mathematical interest, the problem became more

---

[1]Google scholar finds 62.600.000 entries in 0.21 sec for word equations (not all of which is relevant for our topic of course, but narrowing the query down to "word equations" still leads to 1500 entries in 0,16 sec) and several 100,000 more entries if one is patient enough to continue the search and to filter gold from garbage. In the year 2008 at the unification workshop, when we first asked Dr. Google, it found 70.300 entries for word equations in 0.13 sec - so what are we to make of this fact?

widely known, because of its relevance in computer science, artificial intelligence and automated reasoning. Examples are equations over lists with concatenation, data structures such as strings for pattern invoked languages in AI and building equational theories such as associativity into a resolution style theorem prover. Gordon Plotkin [55], Jörg Siekmann [62, 47, 63] and André Lentin [45] independently found an algorithm to compute the set of most general unifiers for the string unification problem, which is infinite in general.

As opposed to the above cited works on decidability, which just enumerate all solutions and make the decidability of the existence of a solution their primary focus, we are more interested in the latter works, inspired by automated theorem proving, where the set $\mu\mathcal{U}\Sigma$ of the *most general* solutions is the focus of attention.

The most common and simple example to show that string unification in free semigroups is infinitary is the following, where a is a constant:

$$(1) \quad xa = ax$$

with the set of most general unifiers

$$\mu\mathcal{U}\Sigma = \{\{x \mapsto a\}, \{x \mapsto aa\}, \{x \mapsto aaa\}, \ldots\}.$$

It is easy to see that indeed this is a solution set and it is not as immediate, but still not too hard to show that there does not exist any other more general set of unifiers $\mu\mathcal{U}\Sigma$ for this problem. Finally $\mu\mathcal{U}\Sigma$ is minimal, which again is obvious, as a^n is always a ground term and thus the unifiers do not yield to instantiation. Hence in general

string unification is infinitary.

As we have said, this is a well known fact since the mid seventies of the last century and it is probably the most often quoted example in any lecture or monograph on unification theory. A similar example

$$(2) \quad xa = bx$$

is usually chosen to demonstrate that the naive string unification algorithms as for example in [55, 62, 47, 63] are not decision procedures: although it is obvious that the above example is not unifiable, the known algorithms would run forever.

In contrast to string unification as it has been understood up to now, problem (1) has a finite set (in fact an even e-unitary set) of essential unifiers

$$e\mathcal{U}\Sigma = \{\{x \mapsto a\}\} = \{\sigma_1\}$$

and any other unifier can be obtained with $\lambda_n = \{x \mapsto a^{n-1}x\}, n > 0$ and $\lambda_2 = \varepsilon$, where ε is the identity substitution.

In other words, for any unifier $\sigma_n = \{x \mapsto a^n\}, n > 1$:

$$\begin{aligned} \sigma_n &= \lambda_n \sigma_1 \\ &= \{x \mapsto a^{n-1}x\} \circ \sigma_1 \\ &= \{x \mapsto a^{n-1}x\} \circ \{x \mapsto a\} \\ &= \{x \mapsto a^n\} \end{aligned}$$

where λ_n obeys a certain structural property, to be defined in the next section.

Once this observation had been made many years ago, there was an intense struggle to find the correct definitions generalizing this observation to the string unification problem and to prove the conjecture[2]

string unification is e-finitary.

As we shall show in this paper, this conjecture is false in general, albeit it holds for certain subclasses of strings.

2 Basic Notions and Notation

Notation and basic definitions in unification theory are well known and have found their way into many and diverse research areas. In particular the monographs and textbooks on automated reasoning always contain sections on unification; most recent research results are presented at the Unification Workshop.[3]

Nevertheless we present the standard notation below - polished for our purpose to see the analogy - followed in contrast by the definitions for our novel approach to essential unification.

2.1 Unification theory: common definitions

A *signature* is a finite set F of function symbols with nonnegative integers, called *arity*, such that an integer n is assigned to each member f of F and f is said to

---

[2]This paper was first published as a preliminary version at the unification workshop in July 2008 [30]. But for various personal problems we submitted (and resubmitted) an expanded and by now completely rewritten version only in 2013 and now finally in 2015.

[3]First workshop in Val d'Ajol in 1987 and since then annually. Since 1997, there is a website UNIF'97, UNIF'98, UNIF'99 up to UNIF'05 in Japan and UNIF'06 at the FLOC conference in Seattle, UNIF'07 and finally UNIF08 at the Schloss Hagenberg, Linz, Austria where a preliminary version of this paper was first presented (see [30]). The current UNIF's can be found at UNIF'13,UNIF'14 and UNIF'15.

be an n-ary function symbol. The subset of n-ary function symbols in F is denoted by F_n. An *algebra* of type F is an ordered pair $\langle A, F \rangle$, where A is a nonempty set and F is a family of finitary operations on A indexed by the signature F such that corresponding to each n-ary function symbol f in F_n there is an n-ary operation f^A on A. The set A is called the universe of the algebra $\langle A, F \rangle$.

Let in the following X be a set of variables and let F be a signature. The set $T(F, X)$ of (syntactic) *terms* of F over X is the smallest set

- comprising X and F_0 and

- if $t_1, \ldots, t_n$ in $T(F, X)$ and f in F_n then the $f(t_1, \ldots, t_n)$ in $T(F, X)$

The set of variable-free terms are called *ground terms*. The set of variables occurring in a term t is denoted by $\mathbf{Var}(t)$. The set of *subterms* of a term $f(t_1, \ldots, t_n)$ contains the term itself and is closed recursively by containing $t_1, \ldots, t_n$ and the sub terms of $t_1, \ldots, t_n$. It is denoted by $\mathbf{Sub}(t)$.

Given F and X, the term *term algebra* of type F over X, denoted by $\langle T(F, X), F \rangle$, has as its universe the set of terms $T(F, X)$ and the fundamental operations satisfying

$$f^{\langle T(F,X),F \rangle}(t_1, \ldots, t_n) = f(t_1, \ldots, t_n)$$

for f in F_n and terms $t_1, \ldots, t_n$ in $T(F, X)$.

A *substitution* is the (unique) homomorphism in the term algebra generated by a mapping $\sigma : X \longrightarrow \mathcal{T}_{F,X}$, which maps a finite set of variables to terms. A substitution σ is represented explicitly as a function by a set of *variable bindings* $\sigma = \{x_1 \mapsto s_1, \ldots, x_m \mapsto s_m\}$. Substitutions are generally denoted by small Greek letters $\alpha, \beta, \gamma, \sigma$ etc. The application of the substitution σ to a term t, denoted $t\sigma$, is defined by induction on the structure of terms

$$t\sigma = \begin{cases} s_i & \text{if } t = x_i \\ f(t_1\sigma, \ldots, t_n\sigma) & \text{if } t = f(t_1, \ldots, t_n) \\ t & \text{otherwise} \end{cases}$$

The substitution $\varepsilon = \{\}$ with $t\varepsilon = t$ for all terms t in $\mathcal{T}_{F,X}$ is called the *identity*. A substitution $\sigma = \{x_1 \mapsto s_1, \ldots, x_m \mapsto s_m\}$ has the *domain*

$$\mathbf{Dom}(\sigma) := \{x \mid x\sigma \neq x\} = \{x_1, \ldots, x_m\};$$

and the *range* is the set of terms

$$\mathbf{Ran}(\sigma) := \bigcup_{x \in \mathbf{Dom}(\sigma)} \{x\sigma\} = \{s_1, \ldots, s_{m'}\}, \ m' \leq m.$$

The set of variables occurring in the range is $\mathbf{VRan}(\sigma) := \mathbf{Var}(\mathbf{Ran}(\sigma))$ and $\mathbf{Var}(\sigma) = \mathbf{Dom}(\sigma) \cup \mathbf{VRan}(\sigma)$. The *restriction* of a substitution σ to a set of variables $Y \subseteq X$, denoted by $\sigma_{|Y}$, is the substitution which is equal to the identity everywhere except over $Y \cap \mathbf{Dom}(\sigma)$, where it is equal to σ. The *composition* of two substitutions σ and θ is written $\sigma \circ \theta$ (to emphasis the composition) or just as $\sigma\theta$ and its application is defined by $t\sigma\theta = (t\sigma)\theta$. This is fine if $\sigma\theta$ has no contradictory variable bindings, otherwise if $x\sigma \neq x\theta$ for some variable x, this binding in θ is applied to σ and eliminated in $\sigma\theta$, (see [7] p 451, for details). A substitution σ is *idempotent* if $\sigma\sigma = \sigma$ and this is true iff $\mathbf{Dom}(\sigma) \cap \mathbf{VRan}(\sigma) = \emptyset$. The application of a substitution to a term can be tricky, if it is not idempotent, e.g. with infinite cycles or contradictory bindings, and there are several solutions proposed for this problem in the literature. In the area of automated reasoning there is the convention that the variables in s_i are always renamed into new variables and contradictory bindings are removed. If σ is not idempotent, then the set representation of a substitution is inadequate, as the application order of the individual bindings matters. In that case $\sigma = \{x_1 \mapsto s_1, x_2 \mapsto s_2, ..., x \mapsto s_m\}$, is often rewritten into "triangle form"[7]:

$$\{x_1 \mapsto s_1\}\{x_2 \mapsto s_2\}....\{x_m \mapsto s_m\}$$

and then applied sequentially and component wise.

Relations such as $=, \geqslant, ...$ between substitutions sometimes hold only if they are restricted to a certain set of variables V. A relation R which is restricted to V is denoted as R^V, and defined as $\sigma\ R^V\ \tau \iff x\sigma\ R\ x\tau$ for all x in V. Two substitutions σ and θ are *equal*, denoted $\sigma = \theta$ iff $x\sigma = x\theta$ for every variable x, they are *equal restricted to V*, $x\sigma =^V x\theta$, iff $x\sigma = x\theta$ for all variables x in V. A term t is an *instance* of a term s denoted $s \leqslant t$, if $t = s\sigma$ for some substitution σ, i.e.

$$s \leqslant t \Leftrightarrow \exists \sigma : s\sigma = t.$$

We also say s is more general or less specific than t, if t is an instance of s. An example is: $s = f(x,y)$, $t = f(x,g(a,b))$ and $\sigma = \{y \mapsto g(a,b)\}$ with $s\sigma = f(x,y)\{y \mapsto g(a,b)\} = t$. There is a little controversy in the literature on wether we should write s$\leqslant$t or t$\leqslant$s: the latter indicates better that s is more general. We prefer the former convention as t has usually more symbols than s. The relation $\leqslant$ is a quasi-ordering on terms called the *instantiation ordering*, or as we prefer to call it a *subsumption ordering*, whose associated equivalence relation and strict ordering are called instantiation (subsumption) equivalence and strict instantiation (subsumption), respectively.

The *encompassment ordering* or *containment ordering* [33, 14, 15] which is our central notion to generalize most general unifiers to essential unifiers is defined as the composition of the sub-term ordering with the instantiation ordering, i.e. a sub-term of t is an instance of s, where $\mathbf{Sub}(t)$ are the sub-terms of t:

$$s \sqsubseteq t \iff \exists \sigma : s\sigma \in \mathbf{Sub}(t).$$

Encompassment conveys the notion that s "appears" in t with a context "above" and a substitution "below". We say t *encompasses* s (s is encompassed by t) or s *is part of* t.

An example is: $s = g(x, b)$ and $t = f(x, g(a, b))$ as above and $\sigma = \{x \to a\}$, because $s\sigma = g(x, b)\{x \to a\} = g(a, b) \in \mathbf{Sub}(t)$.

A substitution σ is called *more general* than θ with $V = \mathbf{Dom}(\theta)$, denoted $\sigma \leqslant^V \theta$, if there exists a λ such that $\theta =^V \sigma\lambda$, i.e.

$$\sigma \leqslant^V \theta \iff \exists \lambda : \theta =^V \sigma\lambda.$$

We also say σ subsumes θ. The relation $\leqslant$, resp. $\leqslant^V$ is a quasi-order, called the *subsumption* ordering for substitutions.

An *equation* or *identity* $s = t$ in a term algebra $\mathcal{T}_{F,X}$ is a pair (s, t) of terms and an algebra A that satisfies the equation $s = t$ if for every homomorphism

$$h : \mathcal{T}_{F,X} \longrightarrow A,$$

$h(s) = h(t)$, that is, only if (s, t) is in the kernel of every homomorphism from $\mathcal{T}_{F,X}$ to A.

An *equational theory* is defined by a set of identities $E \subseteq \mathcal{T}_{F,X} \times \mathcal{T}_{F,X}$. It is the least congruence on the term algebra which is closed under substitution and contains E, and will be denoted by $=_E$. If $s =_E t$ we say s and t are *equal modulo E*. The sets $[s]_E = \{t | t =_E s\}$ are called congruence classes or *equivalence classes* (modulo E).

Definition 1. *Sub-term and instance modulo E*
1. A *sub-term relation* for terms s and t is defined as $s \trianglelefteq t$ iff $s \in \mathbf{Sub}(t)$.
2. A *sub-term relation modulo E* for terms s and t is defined as $s \trianglelefteq_E t$ iff there is a term t' with $t' =_E t$ such that $s \trianglelefteq t'$.
3. A term s is an *instance* modulo E of a term t iff $s\sigma =_E t$.

Definition 2. *Encompassment modulo E*
Term t encompasses term s modulo equational theory E, $s \sqsubseteq_E t$ iff there is a substitution σ such that $s\sigma \trianglelefteq_E t$.
$\sqsubseteq_E$ is the *encompassment relation modulo E*.
Furthermore we say t *strictly encompasses* s, $s \sqsubset_E t$ iff $s \sqsubseteq_E t$, $s \not\sqsupseteq t$.

Let E be an equational theory and Σ be the signature of the underlying term algebra. An *E-unification problem* (over Σ) is a finite set of equations

$$\Gamma = \{s_1 =_E^? t_1, \ldots, s_n =_E^? t_n\}$$

between Σ-terms with variables in a (countably infinite) set of variables, but only a finite set of constants and function symbols in Σ. Let $V = \mathbf{Var}(\Gamma)$.

An *E-unifier of* Γ is a substitution σ, such that

$$s_1\sigma =_E t_1\sigma, \ldots, s_n\sigma =_E t_n\sigma.$$

The set of all E-unifiers of Γ is denoted by $\mathcal{U}\Sigma_E(\Gamma)$ or if the signature Σ is known from the context, we just write $\mathcal{U}_E(\Gamma)$ or even $\mathcal{U}(\Gamma)$. A substitution θ is an instance modulo E of a substitution σ, $\sigma \leqslant_E^V \theta$, iff there exists a λ with $\theta =_E^V \sigma \circ \lambda$. A *complete set of E-unifiers* for Γ is a set C of substitutions, such that

1. $C \subseteq \mathcal{U}\Sigma_E(\Gamma)$, i.e. each element of C is an E-unifier of Γ relative to a signature Σ and

2. for each $\theta \in \mathcal{U}\Sigma_E(\Gamma)$ there exists $\sigma \in C$ with $\sigma \leqslant_E^V \theta$.

The set $\mu\mathcal{U}\Sigma_E(\Gamma)$ is called a *minimal complete set of E-unifiers* for Γ, if it is a complete set, i.e. $\mu\mathcal{U}\Sigma_E \subseteq C$, and there are no elements σ, σ' in $\mu\mathcal{U}\Sigma_E$ with $\sigma <_E^V \sigma'$, i.e. $\sigma \leqslant_E^V \sigma'$ implies $\sigma =_E^V \sigma'$ for all $\sigma, \sigma' \in \mu\mathcal{U}\Sigma_E$. When a minimal complete set of E-unifiers of a unification problem Γ exists, it is unique up to $=_E^V$.

The empty or unit substitution ε is a unifier for s and t in case $s =_E t$. Minimal complete sets of unifiers need not always exist, and if they do, they might be singular, finite, or infinite. Since minimal complete sets of E-unifiers are isomorphic whenever they exist they can be used to classify theories with respect to their corresponding unification problem as well.

This leads naturally to the concept of a *unification hierarchy* which was first introduced in Siekmann's doctoral thesis in 1976 [63] and further refined and extended by himself and his later students as well as by many subsequent workers in the field of unification theory, see [65, 42, 6, 24, 7] for standard surveys.

A *unification problem* Γ is *nullary*, if for a solvable unification problem Γ the set of $\mu\mathcal{U}\Sigma_E(\Gamma)$ does not exist. The unification problem Γ is *unitary*, if it is not nullary and the minimal complete set of E-unifiers for Γ is of cardinality less or equal to 1. The unification problem Γ is *finitary*, if it is not nullary and the minimal complete set of E-unifiers is of finite cardinality. The unification problem Γ is *infinitary*, if it is not nullary and the minimal complete set of E-unifiers is infinite.

An *equational theory* E is *unitary*, if all unification problems for E are unitary. An equational theory E is *finitary*, if all unification problems are finitary. An equational theory E is *infinitary*, if there is at least an infinitary unification problem and all unification problems have minimal complete sets of E-unifiers. If there exists a unification problem Γ not having a minimal complete set of E-unifiers, then the equational theory E is *nullary* or of *type zero*.

2.2 Additional Definitions: Essential E-unifiers

Substitutions form a semigroup with respect to their composition. This fact was used to define the subsumption order on unifiers from above, namely

$$\sigma \leq_E^V \tau \iff \exists \lambda : \tau =_E^V \sigma \circ \lambda,$$

where $V = \mathbf{Dom}(\tau)$, which led to the notion of a most general unifier.

As argued above this concept does not generalize well on equational theories: the equational theory of associativity $A = \{x(yz) = (xy)z\}$, i.e. the free semigroup with the unification problem $\Gamma = \{ax =_A^? xa\}$ has the infinite set of most general unifiers $\{\{x \mapsto a^n\} | n \geq 1\}$, as discussed in the introduction. However, the essential unifier in this set intuitively seems to be $\sigma = \{x \mapsto a\}$, because every most general unifier contains this unifier in a certain sense, namely let:

$$\tau_n = \{x \mapsto a^n\} = \{x \mapsto a^{n-1}x\} \circ \sigma.$$

Now having in mind that substitutions form a semigroup, the dual of the instantiation ordering, i.e. left-composition instead of right-composition changes the infinitary problem into a finitary one, because if we redefine the order $\leq_A$ into $\exists \lambda : \tau =_A \lambda \sigma$, where $\sigma = \{x \mapsto a\}$ and for $\tau_n = \{x \mapsto a^n\}$ we then have $\lambda_n = \{x \mapsto a^{n-1}x\}$. But this is not compatible with the original notion of generality and it would not quite work in general.

Our solution is therefor based on lifting the encompassment order on terms to an *encompassment order on substitutions* (modulo an equational theory E). For this we use the fact, that any substitution τ can be decomposed into three parts, $\tau =_E \lambda_1 \sigma \lambda_2$, the most trivial decomposition would be the one with an identity substitution on both sides, i.e. $\tau =_E \varepsilon \tau \varepsilon$. Viewing the substitution τ that way, we can say that an instance of σ, namely $\sigma \lambda_2$ is a "sub-part" of τ and we require also that $\mathbf{Dom}(\tau) = \mathbf{Dom}(\sigma)$. This observation allows the definition of the concept, that τ encompasses σ (modulo E) in the following way:
We defined in section 2.1 Definition 1 (2) for an equational theory E, that a term s is encompassed by the term t, $s \sqsubseteq_E t$, if there exists a substitution σ with $s\sigma \trianglelefteq_E t$.

In analogy we define $\sigma \sqsubseteq_E \tau$ iff $\exists \lambda_1$ and $\exists \lambda_2$ such that $\mathbf{Dom}(\tau) = \mathbf{Dom}(\sigma) =: V$ and $\tau =_E^V \lambda_1 \sigma \lambda_2$, i.e. for all x from $\mathbf{Dom}(\tau)$: $x(\tau) =_E x(\lambda_1 \sigma \lambda_2)\,|_V$. We say σ is *encompassed by τ modulo E* or *τ encompasses σ modulo E*.

So let us cast all this into formal definitions.

Definition 3. A substitution σ is a *sub-substitution* of τ iff $\mathbf{Dom}(\sigma) = \mathbf{Dom}(\tau)$ and for all $x \in \mathbf{Dom}(\sigma)$ we have $x\sigma \in \mathbf{Sub}(x\tau)$. Let $\mathbf{SUB}(\tau)$ be the set of all sub-substitutions of τ.

For example if $\tau = \{x \mapsto f(a,z)\}$, then

$$\mathbf{SUB}(\tau) = \{\ \{x \mapsto a\},$$
$$\{x \mapsto z\},$$
$$\{x \mapsto f(a,z)\}\ \}$$

because $x\{x \mapsto a\} = a \in \mathbf{Sub}(x\tau) = \mathbf{Sub}(f(a,z)) = \{f(a,z), a, z\}$ and similarly for the other components.

Definition 4. *encompassment order for substitutions*

A substitution σ is *encompassed modulo E* by a substitution τ, or *τ encompasses σ modulo E*, $\sigma \sqsubseteq_E \tau$, iff $\mathbf{Dom}(\tau) = \mathbf{Dom}(\sigma) = V$ and there exists a substitution λ and a substitution τ' with $\tau' =_E \tau$ such that $(\sigma\lambda)\,|_V \in \mathbf{SUB}(\tau')$, in other words $(\sigma\lambda)$ restricted to V is a sub-substitution of τ modulo E.

$\sigma \sqsubset_E \tau$ denotes *strict encompassment modulo E*.

In order to see the analogy to the encompassment definition for terms, consider the following two terms s and t in analogy to the two substitutions σ and τ:

$$s = f(x,y),\ t = f(x,g(a,b)),\ \sigma = \{y \to g(a,b)\}$$

then $s \notin \mathbf{Sub}(t)$, but $s\sigma \in \mathbf{Sub}(t)$, i.e. $s \sqsubset t$.

Now consider the substitutions τ and σ with:

$$\tau = \{x \mapsto f(a,b),\ y \mapsto f(a,g(a,b))\}$$
$$\sigma = \{x \mapsto a,\ y \mapsto g(a,z)\}$$
$$\lambda_2 = \{z \mapsto b\}$$

then $\sigma \notin \mathbf{SUB}(\tau)$ but $(\sigma\lambda_2)\,|_V \in \mathbf{SUB}(\tau)$, i.e. $\sigma \sqsubset \tau$ and with

$$\lambda_1 = \{\,x \mapsto f(x,b),\ y \mapsto f(a,y)\}$$

we have a tripartition of τ:

$$\begin{aligned}\tau =^V \lambda_1\sigma\lambda_2 &=^V \{x \mapsto f(x,b),\, y \mapsto f(a,y)\}\{x \mapsto a,\, y \mapsto g(a,z)\}\{z \mapsto b\} \\ &=^V \{x \mapsto f(x,b),\, y \mapsto f(a,y)\}\{x \mapsto a,\, y \mapsto g(a,b)\} \\ &= \{x \mapsto f(a,b),\, y \mapsto f(a,g(a,b))\}\end{aligned}$$

If a substitution is a unifying substitution (for an E-unification problem Γ), we define:

Definition 5. *part unifiers, essential unifiers*

1. An E-unifier σ for a unification problem Γ modulo the equational theory E and the variables $V = \mathbf{Var}(\Gamma)$, is *encompassed* by an E-unifier τ for Γ, denoted as above by $\sigma \sqsubseteq^V_E \tau$, if there exists a substitution λ, such that $(\sigma\lambda)|_V$ is a sub-substitution of τ.

2. We say τ contains a *part unifier* σ iff τ encompasses σ **and** σ is an E-unifier as well.

3. An E-unifier σ for a unification problem Γ modulo the equational theory E that does not encompass any other E-unifier for Γ is called an *essential E-unifier*. We denote the set of essential E-unifiers as $e\mathcal{U}\Sigma_E(\Gamma)$. Two unifiers σ and τ are *encompassment equivalent modulo E*, denoted $\approx^V_E$, if $\sigma \sqsubseteq^V_E \tau$ and $\tau \sqsubseteq^V_E \sigma$.

4. A *complete* set of *essential E-unifiers* for Γ is a set of E-unifiers, such that for each E-unifier τ there exists σ in the set with $\sigma \sqsubseteq^V_E \tau$.

5. The set $e\mathcal{U}\Sigma_E(\Gamma)$ is called a *minimal complete set* of *essential E-unifiers* for Γ, or simply *the set of essential E-unifiers for Γ*, if it is a complete set and for all σ and σ' in $e\mathcal{U}\Sigma_E(\Gamma)$ σ and σ' are encompassment equivalent.

Proposition 6. *The encompassment order on substitutions is a quasi order, i.e. reflexive and transitive.*

Proof. reflexivity: $\sigma \sqsubseteq_E \sigma$ means there are substitutions $\lambda_1, \lambda_2 : \sigma =_E \lambda_1\sigma\lambda_2$, setting λ_1 and λ_2 to the substitution identity ε leads to $\sigma =_E \varepsilon\sigma\varepsilon = \sigma$.

transitivity: $\sigma \sqsubseteq^V_E \tau$ and $\tau \sqsubseteq^V_E \psi$ implies $\sigma \sqsubseteq^V_E \psi$, where by definition we have $\mathbf{Dom}(\sigma) = \mathbf{Dom}(\tau) = \mathbf{Dom}(\psi) =: V$, so

$$\begin{aligned}\tau &=^V_E \lambda_{1,1}\sigma\lambda_{2,1} \\ \psi &=^V_E \lambda_{1,2}\tau\lambda_{2,2}\end{aligned}$$

which implies
$$\psi =_E^V (\lambda_{1,2}\lambda_{1,1}\sigma\lambda_{2,2}\lambda_{2,1}) \Rightarrow \sigma \sqsubseteq_E^V \psi$$

□

A set of unifiers $C(\Gamma)$ is *e-complete* for Γ if for every unifier σ there exists a unifier τ in C which is a part unifier of σ. A complete set of unifiers $C(\Gamma)$ is *e-minimal* if any two distinct elements are not part of each other. This set is denoted as $e\mathcal{U}\Sigma_E(\Gamma)$ and it is unique up to part equivalence. Because if it would not be unique there would exist two complete sets of essential unifiers $e\mathcal{U}\Sigma_E^1$ and $e\mathcal{U}\Sigma_E^2$ with τ in $e\mathcal{U}\Sigma_E^1 \setminus e\mathcal{U}\Sigma_E^2$ and σ in $e\mathcal{U}\Sigma_E^2 \setminus e\mathcal{U}\Sigma_E^1$. But since $e\mathcal{U}\Sigma_E^1$ is complete, $\tau \sqsubseteq_E \sigma$ and since $e\mathcal{U}\Sigma_E^2$ is a set of essential unifiers there is a σ' in $e\mathcal{U}\Sigma_E^2 \setminus e\mathcal{U}\Sigma_E^1$ with $\sigma' \sqsubseteq_E \tau$. But then σ and σ' are part equivalent. Therefore $\sigma \sqsubseteq_E \tau$, which means σ and τ are part equivalent, contradicting the assumption.

It can be shown that the set of essential unifiers $e\mathcal{U}\Sigma_E(\Gamma)$ can be used to generate all unifiers, just as the set of most general unifiers is used to generate all unifiers.

Lemma 7. *Let Γ be a non nullary E-unification problem, then $e\mathcal{U}\Sigma_E(\Gamma) \subseteq \mu\mathcal{U}\Sigma_E(\Gamma)$, i.e. the set of essential unifiers is always a subset of the set of most general unifiers.*

Proof. If σ is an essential E-unifier it has by definition no part unifier, i.e. σ is not an instance of any other unifier, hence σ is also a most general unifier. Essentials are usually a proper subset, since there are equational theories, like string unification, with most general unifiers σ, τ and $\tau =_E \lambda_1\sigma$, so τ is not an essential. For example $\{ax =_A^? xa\}, \sigma = \{x \mapsto a\}, \tau = \{x \mapsto aa\}$ and with $\lambda_1 = \{x \mapsto ax\}$ we have $\tau =_E \lambda_1\sigma$. □

Obviously, an essential unifier of a nunnullary problem can not be the instance of another unifier, therefore it is most general. But a most general unifier could contain essential unifiers.

The interesting observation is that the above subset of essential E-unifiers can be lovely, i. e. extremely small in comparison to its superset, as we shall see in the following.

3 Essential String Unification

We are interested now in A-unification, i.e. unification in a *free semigroup*, where

$$A = \{f(x, f(y, z)) = f(f(x, y), z)\}$$

and the set of terms are built up as usual over constants, variables, but only one function symbol f. In this case, we can just drop the fs and brackets and write the sequence of symbols as strings (or words) over the alphabet of constants and variables. The *empty string* (a string with length 0) will be denoted as ε and it is the identity element, which turns a semi group into a *monoid*. In the following we simply write $=$ for the equational sign $=_A$. A set of string equations will be denoted as $\Gamma = \{u_1 = v_1, \ldots, u_n = v_n\}$ where the u_i and the v_i are strings (words). $Var(\Gamma)$ is the set of free variable symbols occurring in u_i and v_i. Let $V = Var(\Gamma)$, then a *(string-)unifier* $\sigma : V \to (\Sigma \backslash V)^*$ is a solution for Γ if $u_i \sigma = v_i \sigma, 1 \leqslant i \leqslant n$. The set of all unifiers is denoted as $\mathcal{U}\Sigma_A(\Gamma)$ and we drop the A when it is clear from the context. A unifier σ is *ground* if its range contains only constants and no variables.

Now let us look at a few motivating examples, which show that indeed an infinite set of most general unifiers $\mu\mathcal{U}\Sigma$ collapses to a finite set of **essential unifiers** $e\mathcal{U}\Sigma$, supporting the earlier hypothesis that the infinitary string unification problem is essentially finitary.

Our first example is the well known string unification problem mentioned in the introduction:
$$ax =^? xa \text{ with } \sigma_n = \{x \mapsto a^n\}, n > 0$$
has infinitely many most general unifiers σ_n, but there is just *one* e-unifier $\sigma_1 = \{x \mapsto a\}$ because of
$$\sigma_n = \{x \mapsto a^{n-1}x\} \circ \sigma_1.$$

The next example has two variables[4]
$$xy =^? yx$$
and it has infinitely many most general unifiers
$$\sigma_{i,j} = \{x \mapsto z^i, y \mapsto z^j\}, i, j > 0,$$
where i and j are relative prime, i.e. $gcd(i,j) = 1$.
But it has only one e-unifier $\sigma_1 = \{x \mapsto z, y \mapsto z\}$ because of
$$\sigma_{i,j} = \{x \mapsto z^{i-1}x, y \mapsto z^{j-1}y\} \circ \sigma_1$$

Our next example is taken from J. Karhumäki in *Combinatorics of Words* [12] see also [48, 49]. The system
$$\left\{ \begin{array}{l} xaba =^? baby \\ abax =^? ybab \end{array} \right\}$$

---
[4] see http://www.math.uwaterloo.ca/~snburris/htdocs/WWW/PDF/e_unif.pdf, example 15

has infinitely many most general unifiers

$$\sigma_n = \{x \mapsto b(ab)^n, y \mapsto (ab)^n a\}, n \geq 0$$

But it has only one e-unifier $\sigma_0 = \{x \mapsto b, y \mapsto a\}$ because of

$$\sigma_n = \{x \mapsto x(ab)^n, y \mapsto (ab)^n y\} \circ \sigma_0.$$

Exploiting the analogy between the first and the second example above, we can easily construct more examples in this spirit.

Our fourth example is taken from J. Karhumäki as well:

$$axxby =^? xaybx$$

It has infinitely many most general unifiers

$$\sigma_{i,j} = \{x \mapsto a^i, y \mapsto (a^i b)^j a^i\}, i \geq 1, j \geq 0$$

but it has only one e-unifier $\sigma_{1,0} = \{x \mapsto a, y \mapsto a\}$ which is essential because of

$$\sigma_{i,j} = \{x \mapsto ya^{i-1}, y \mapsto (a^i b)^j xa^{i-1}\} \circ \sigma_{1,0}$$

i.e. $\sigma_{1,0} \sqsubseteq \sigma_{i,j}$ and $\sigma_{1,0}$ has no part unifier. The final example is a bit more elaborate but still in the same spirit.

$$zaxzbzy =^? yyzbzaz$$

has infinitely many most general unifiers

$$\sigma_n = \{x \mapsto b^{2n}a, y \mapsto b^n ab^n, z \mapsto b^n\}, n \geq 1$$

but it has only one e-unifier, namely $\sigma_1 = \{x \mapsto bba, y \mapsto bab, z \mapsto b\}$ because of

$$\sigma_n = \{x \mapsto b^{2n-2}x, y \mapsto b^{n-1}yb^{n-1}, z \mapsto b^{n-1}z\} \circ \sigma_1$$

3.1 String Unification with at most one variable is e-finitary

Let us assume that our unification problem

$$\Gamma = \{u_1 =^? v_1, \ldots, u_n =^? v_n\}$$

over the signature $\Sigma = Con \cup X$ consists of at most one variable from X, but arbitrary many constants from Con. Without loss of generality, each set of string equations can be encoded into a single string equation preserving the solutions, which is well known (for example see J.I. Hmeleyskij [29]). Volker Diekert in [16] used the following construction

$$\{u_1 a \ldots u_n a u_1 b \ldots u_n b =^? v_1 a \ldots v_n a v_1 b \ldots v_n b\}$$

where a and b are distinct constants. It can be shown, that the two equational problems have the same solutions. We shall also use results from [13, 53], which show in Theorem 3 of [13] that a word equation in at most one variable has either (i) no solution or (ii) finitely many solutions bounded by $\mathcal{O}(log \ | \ \Gamma \ |)$ or (iii) one infinite solution of the form $(uv)^+ u$ for some words u and v. Using these facts we have that the string unification problem in one variable is of type infinitary, which was already shown by the first case $xa =^? ax$ above and we even have the stronger result:

Theorem 8. *A string unification problem Γ in one variable has either no solution or the minimal and complete set is $\mu\mathcal{U}\Sigma_A(\Gamma) = F \cup \{x \to (pq)^{i+1} p, i \geqslant 0\}$ for some p, q in Σ, where pq is primitive and F is a finite set of unifiers whose number is bounded by $\mathcal{O}(log \ | \ \Gamma \ |)$.*

Proof. Follows easily from the proof of theorem 3 in [13] □

Note: A word is primitive if it is not the power of another word.

Corollary 9. *Let $\Gamma = \{u_0 x u_1 \ldots x u_n = v_0 x v_1 \ldots x v_m\}$, where u_i, v_i are ground strings and $\mathbf{Var}(\Gamma) = V = \{x\}$.*

9.1 The equation in Γ can be reduced to the form $\Gamma' = \{u_0 x u_1 \ldots x u_n = x v_1 \ldots x v_m\}$, where u_0 is not the empty string and either u_n is nonempty and v_m is empty or vice versa. This form implies also that for any unifier $\sigma = \{x \mapsto w\}$, with a ground string $w \in Con^+$ $x\sigma$ is a prefix of the string u_0^k.

9.2 If $m \neq n$ there is at most one unifier.

9.3 If $m = n = 1$, i.e. $\Gamma' = \{u_0 x = x v_1\}$, then the unifiers are of the form: $\{x \mapsto (pq)^i p, i \geqslant 0\}$, where pq is primitive.

9.4 If $m = n > 1$ the unifiers are of the form: $\{x \mapsto (pq)^{i+1} p, i \geqslant 0\} \cup F$, as in Theorem 8 above.

9.5 For a given Γ there exists at most one infinite solution of the form: $\sigma_i = \{x \mapsto (pq)^{i+1} p\}, i \geqslant 0$ see Lemma 1 in [13].

These known results (taken inter alia from [13] and [53], and for a recent result see also [37]) imply that the string unification problem in one variable is of type infinitary, as demonstrated by the $ax =^? xa$ case above. Using some of their results,

we will now show that *string unification with only one variable* is of type *e-**finitary***.
The first step is to prove that all unifiers are ground substitutions. The second step
is to prove that there are at most finitely many essential unifiers.

Proposition 10. *Let* $\Gamma = \{u_0 x u_1 ... x u_n =^? x v_1 ... x v_n\}$ *be a solvable string equation with different input terms and with at most one variable x. Then all unifiers are ground substitutions, i.e.* $\forall \sigma \in \mathcal{U}\Sigma_A(\Gamma) : x\sigma \in Con^*$

Proof. Suppose by contradiction with an arbitrary unifier $\{x \mapsto w\} \in \mathcal{U}\Sigma_A(\Gamma) : w = w_1 z w_2$ where z is a new variable $z \neq x$ such that w_1 is the ground prefix of w. Applying the unifier $x \mapsto w_1 z w_2$ yields

$$u_0 w u_1 ... = w v_1 ... = u_0 w_1 z w_2 u_1 ... = w_1 z w_2 v_1 ...$$

Consider the prefixes $u_0 w_1 ... = w_1 z ...$ Since $|u_0 w_1| \geq |w_1 z|$ and u_0 is nonempty, z must be a symbol in $u_0 w_1$, which is impossible, since u_0 and w_1 are ground by assumption.

Hence **Var**(w) is empty and $\mathcal{U}\Sigma_A(\Gamma) = \mu\mathcal{U}\Sigma_A(\Gamma)$. □

Theorem 11. *String unification with one variable is e-finitary and the number of unifiers is bounded by* $\mathcal{O}(\log |\Gamma|)$.

Proof. By Theorem 8 above the nonessential solution set consists of at most finitely many unifiers (in F) and at most one infinite set of unifiers of the form
$\{x \mapsto (pq)^{i+1}p, i \geq 0\}$ for some p, q where pq is primitive.
So let us look at the infinite set and the proof is carried out by cases.
In **case** $m \neq n$ then by proposition 9.2 there is only one unifier, which is hence in particular essential.
In **case** $m = n = 1$ then by proposition 9.3, the unifiers are of the form
$\sigma_i = \{x \mapsto (pq)^i p\}, i \geq 0$ where pq is primitive.
Now p is either empty or not.
- If p is empty then $\sigma_i = \{x \mapsto q^i\}$, $i \geq 1$. But then $\sigma_1 = \{x \mapsto q\}$ is the single essential unifier, because with $\lambda = \{x \mapsto q^{i-1}x\}$ we have
$\sigma_i = \lambda \sigma_1 = \{x \mapsto q^{i-1}x\} \circ \{x \mapsto q\}$ for $i \geq 1$.
- If p is nonempty then $\sigma_i = \{x \mapsto (pq)^i p\}, i \geq 0$ and $\sigma_0 = \{x \mapsto p\}$ is the only essential unifier, since with $\lambda = \{x \mapsto (pq)^i x\}$ we have
$\sigma_i = \{x \mapsto (pq)^i x\} \circ \{x \mapsto p\} = \{x \mapsto (pq)^i p\}$.
Hence either $\sigma_1 = \{x \mapsto q\}$ or $\sigma_0 = \{x \mapsto p\}$ is the single essential unifier.
In **case** $m = n > 1$ we have by proposition 9.4
$\sigma_i = \{x \mapsto (pq)^{i+1}p\}, i \geq 0$.
Now again if p is empty we have $\sigma_0 = \{x \mapsto q\}$ as the essential unifier, because

$\sigma_i = \{x \mapsto q^{i+1}\} = \{x \mapsto q^i x\} \circ \{x \mapsto q\}$ $i \geq 0$.

If p is nonempty then $\sigma_0 = \{x \mapsto pqp\}$ is the only essential unifier since $\sigma_i = \{x \mapsto (pq)^{i+1}p\} = \{x \mapsto (pq)^i x\} \circ \{x \mapsto pqp\}$, $i > 0$.

Hence either $\{x \mapsto pqp\}$ or $\{x \mapsto q\}$ are essential unifiers.

So in summary the set of solutions can be represented by just one essential unifier for the infinite case. Now since there are at most finitely many solutions in F bounded by $\mathcal{O}(log \mid \Gamma \mid)$, the set of essential unifiers is finite as well, because $e\mathcal{U}\Sigma_A(\Gamma) \subseteq \mu\mathcal{U}\Sigma_A(\Gamma) \subseteq \mathcal{U}\Sigma_A(\Gamma)$. □

3.2 String unification in general is e-infinitary

String unification with at most one variable in the signature Σ is e-finitary as we have seen above and surely there are many more special cases of signature restrictions, where the set of e-unifiers is always finite or even unitary. Special cases of this nature have also been investigated extensively in the past for the solvability problem of word equations in the mathematical community (see for example the IWWERT workshop series) as well as in the automated reasoning communities for string unification (for example at the UNIF workshop series).

However the general result for e-string unification is:

Theorem 12. *String unification with more than one variable is e-infinitary*

Proof. For $\Gamma = \{xby =^? ayayb\}$ the set of essential unifiers is

$$e\mathcal{U}\Sigma_A(\Gamma) = \{\{x \mapsto ab^n a, y \mapsto b^n\} : n > 0\}$$

Correctness
Any substitution $\sigma_n = \{x \mapsto ab^n a, y \mapsto b^n\}$ is a unifier since $(xby)\sigma_n = ab^n abb^n = ab^n ab^{n+1} = (ayayb)\sigma_n$.

Completeness
We show that any unifier is of the form $\{x \mapsto ab^n a, y \mapsto b^n\}$. Consider some unifier $\{x \mapsto u, y \mapsto v\}$. Since $\Gamma = \{xby = ayayb\}$, $u = au'$ and $v = v'b$. Applying the unifier in $xby = ayayb$ yields $au'bv'b = av'bav'bb$. Since v' can not contain any a, $v' = b^i$, and $v = b^{i+1}$, $i \geq 0$. Hence we have $au'b^{i+1} = ab^{i+1}ab^{i+1}b$, which yields $u' = b^{i+1}a$. Hence $u = ab^{i+1}a, v = b^{i+1}, i \geq 0$.

Consequently $\mathcal{U}\Sigma_A(\Gamma) = \{\{x \mapsto ab^n a, y \mapsto b^n\} : n > 0\}$

Essential
We show that the set $\mathcal{U}\Sigma_A(\Gamma)$ is e-minimal. So take any pair of different unifiers $\{x \mapsto ab^m a, y \mapsto b^m\}$ and $\{x \mapsto ab^n a, y \mapsto b^n\}$ and we show that they are incomparable with respect to the encompassment ordering, i.e. no unifier from $\mathcal{U}\Sigma_A(\Gamma)$ has a part unifier.

Suppose from $m < n$ that σ_m is a part unifier of σ_n. From Definition 3 and Definition 4 it follows, that $\sigma_m \sqsubseteq_A \sigma_n$ i.e. $x\sigma_m$ is a substring of $x\sigma_n$. But $x\sigma_m = ab^m a$ is substring of $x\sigma_n = ab^n a$ only if $m = n$, contradicting $m < n$. Hence for $\forall \sigma \in \mathcal{U}\Sigma_A(\Gamma)$ σ is an essential unifier and we have $\mathcal{U}\Sigma_A(\Gamma) = \mu\mathcal{U}\Sigma_A(\Gamma) = e\mathcal{U}\Sigma_A(\Gamma)$. □

3.3 A General A-Theorem

Let E be a set of equational axioms containing the associativity axiom of a binary operator $*$, i.e. $A = \{x * (y * z) = (x * y) * z\}$ and $E = A \cup R$, where R is some set of equations. We call the theory modulo E A-separate, if any equation in R can not be applied to a pure string $s_1 * s_2 * \cdots * s_n$ (the brackets are suppressed).

For instance consider distributivity (which is an infinitary unification theory, see [69, 5]):

$$D = \{x * (y + z) = (x * y) + (x * z), (x + y) * z = (x * z) + (y * z)\},$$

then the theory of $E = A \cup D$ is A-separate. To see this, note that no equation of D can be applied to a string of $x_1 * x_2 * \cdots * x_n$, simply because there are no sums involving the plus sign $+$, but each equation in D has the sum symbol $+$ on its left and on its right hand side.

Formally, $E = A \cup R$ is A-separate, if for all elements u of the A-theory $u =_R v$ implies $u = v$.

Theorem 13. *All not e-nullary A-separate E-theories are e-infinitary*

Proof. Consider the unification problem of section 3.2 above: In the associative subalgebra it has infinitely many e-unifiers. Each of the elements of the range of the essential unifiers is not affected by the remaining equational axioms in $R = E \setminus A$, since E is A-separate. Hence each A-separate theory is e-infinitary. □

As noted above the not e-nullary theory $A \cup D$ is A-separate and hence:

Theorem 14. *The theory $A \cup D$ is e-infinitary.*

Note that the theorem does not imply that D alone is e-infinitary: D is known to be infinitary [69, 68], but the essential case for D has not yet been examined.

4 Idempotent semigroups are e-finitary

The following theory of *idempotent Semigroups or Bands* defined by

$$AI = \{f(x, f(y, z)) = f(f(x, y), z), f(x, x) = x\}$$

demonstrates another interesting case for essential unifiers. Note that this theory is not *A*-separate. The theory is nullary with respect to the instantiation order, since there are solvable *AI*-unification problems which do not posses a minimal complete set of *AI*-unifiers with respect to the instantiation ordering, see [3, 59].

However, with respect to the encompassment ordering $\sqsubseteq_E$ this well-known situation changes completely as this theory is in fact finitary. Associativity and idempotency constitute the algebra of idempotent strings and it has been shown in [31] that:

Proposition 15. *The theory AI is not nullary with respect to essential unifiers.*

There are *AI*-unification problems with more than one essential unifier. Therefore:

Proposition 16. *AI is not unitary with respect to essential unifiers.*

And finally the most striking result:

Theorem 17. *The theory AI is finitary with respect to essential unifiers.*

5 A Derivation System for essential A-Unification

An important requirement for any unification algorithm to be built into an automated reasoning system is that it can *incrementally* generate the set of most general (essential) unifiers without backtracking. Can we do this for essential unifiers as well?

Since the property of a unifier to be essential (or not) is decidable, as we will show in the appendix, we can proceed as follows: we compute incrementally the set of most general unifiers for the two input strings and eliminate each mgu that is not essential as we go. Of course more elaborate techniques are possible, in fact there is a small but active research community of mathematicians concerned with these problems (see [48, 49]). However our main interest here and in an upcoming sequel of contributions is to see how far we can go in general - i.e. not just for strings - with essential unification based on the encompassment relation rather than the usual instantiation ordering.

Currently there are two basic techniques for standard string unification: the early algorithms (as for example in [62, 47, 63]) generate a search tree by parsing the two strings to be unified from left to right taking the property of the leftmost symbol into account. The second approach (as in [26, 66]) is based on the key insight that the two finally unified strings/words must have the same length and hence we can set up a diophantine equation for each variable of the given strings.

We shall use the second approach and show in the appendix how this can be carried out in detail.

6 Conclusion

For the theoretically inclined reader our result is likely to be of interest with respect to the unification hierarchy for essentials: there are obviously essentially unitary E-unification problems (theories); there are unexpected essentially finitary E-unification problems (theories), for example the AI-unification problem, the one variable string unification case of this paper or the essential unification problem for commutativity [?] and finally there are essentially infinitary E-unification problems, which hold some surprises as well, as this paper shows. The question of the existence of an essential nullary E-unification problem is discussed in [?].

But in any case, the amount of most general unifiers is substantially reduced; in other words they are not really *most general* at all when it comes to E-unification. For example the theory of associativity and commutativity AC has exponentially many unifiers, in fact for a base B there are B with a tower of exponentials of size n many unifiers [8, 18, 39, 40]. So the encompassment order may be *the order of choice* for E-unification, rather than the standard instantiation $\leqslant$-order for most general unifiers.

For the reader who is - like us - more interested in practical automated reasoning systems, the results reported here come as a disappointment to some extent: while a finite set of essential unifiers modulo E is considerably smaller than the set of most general unifiers for a string unification problem, the hypotheses that the infinitary A-theory collapses into an e-finitary theory did not hold up to scrutiny.

This may not surprise the reader familiar with this problem: in spite of the simplicity and immediate intuitiveness of the problem formulation (using strings or words) its solvability as well as the unification problem turned out to be of exceptional difficulty and complexity and stayed open in the mathematical community for several decades (see [60, 16]). It motivated inter alia the large amount of work on semigroups and words, which hold far more expressivity and complexity as one may have thought half a century ago (as witnessed for example by the workshop series IWWERT and others, see [48, 49]).

For practical purposes as a unification component within an automated theorem proving system, based on resolution or rewriting, there are two problems that still have to be solved (just as for any other theory):

1. To find a unification algorithm which generates — as efficiently as possible — the set of essential A-unifiers.

2. To show how the reasoning machinery, for example a theorem prover based on resolution, can be built upon essential unifiers instead of most general unifiers,

which is not easy as the standard lifting lemma does not work for essentials and has to be replaced by some other technique.

There is a solution to (1) as presented in this paper that is essentially based on enumeration, and we have a solution to (2) based on straight enumeration as well. However this is far from anything practically useful: the unification algorithm is to resolution based theorem proving what the addition-and-multiplication unit is in a general purpose computer and hence deserves the utmost effort in engineering (see [58] for an early proposal), measured not in MiPs (million instructions per second), but in LiPs (logical inferences per sec, that is in fact the number of unifications per sec) which was the hallmark of the fifth generation computer race in the 1980s.

A A derivation system for essential A-Unification

In the following we spell out the details of an e-unification algorithm for strings. As this procedure is based on more or less well known techniques plus step-by-step elimination, we present these details here in the appendix, as we feel that they nevertheless need to be spelled out in painstaking detail, in order to convince ourselves of its correctness and completeness.

A.1 Linear Diophantine Equations

Let $\Gamma = \{u =_A^? v\}$ be an A-unification problem. The first observation is that any solution maps u and v into words of equal length, i.e for any unifier σ we have $|u\sigma| = |v\sigma|$, where $|u|$ gives the number of symbols in word u. As in Makanin's algorithm [50] and in the AC-algorithm of Herold, Siekmann [26] as well as Stickel [66], we intend to map this string (word) unification problem (for the calculation of the length of possible string-unifiers) into a linear diophantine equation. This is done by interpreting the variables as diophantine variables and the length of symbols as integer coefficients using a homomorphism Λ between the strings of the free monoid $(X \cup \Sigma)^*$ into the integer polynomials $\mathcal{P}(X)$, called the linear "length" Λ :

$$\Lambda : x \mapsto \begin{cases} x \text{ for } x \in \mathbf{Var}(\Gamma) \\ 1 \text{ otherwise} \end{cases} \quad \text{and} \quad \Lambda(uv) = \Lambda(u) + \Lambda(v)$$

where X is the set of variables and Σ is the signature as above. $\mathcal{P}(X)$ are the polynomials in X with integers as coefficients and natural numbers as solutions. The equational theory for $\mathcal{P}(X)$ could be the set of Peano-axioms, denoted as P.

A unification problem $\Gamma = \{u =_A^? v\}$ is then translated into a linear diophantine equation (the Peano theory P) with solutions in $\mathbb{N}$:

$$\Lambda(\Gamma) = \{\Lambda(u) =_P^? \Lambda(v)\}.$$

For instance let Σ be the alphabet $\{a, b\}$ and let $\Gamma = \{xby =_A^? ayayb\}$. Then

$$\Lambda(\Gamma) = \{x + 1 + y =_P^? 1 + y + 1 + y + 1\} = \{x - y =_P^? 2\}.$$

One solution for this equation is for example $x = 4$ and $y = 2$, i.e. x should substitute just 4 new variable symbols and y should substitute just 2 new variable symbols. This requirement can be captured by substituting four new variables x_1, x_2, x_3, x_4 into x and two new variables y_1, y_2 into y with the proviso that x_i and y_i should only have *one* symbol as its value.

Let us also translate a substitution $\sigma = \{x_1 \mapsto u_1, \ldots, x_n \mapsto u_n\}$ into a "diophantine substitution":

$$\Lambda(\sigma) = \{x_1 \mapsto \Lambda(u_1), \ldots, x_n \mapsto \Lambda(u_n)\}.$$

Now for the unifier $\sigma = \{x \mapsto abba, y \mapsto bb\}$ we have

$$\Lambda(\sigma) = \{x \mapsto 4, y \mapsto 2\},$$

which is obviously a solution for $\Lambda(\Gamma)$. As mentioned above, any unifier σ for Γ maps into words of equal length $|u\sigma| = |v\sigma|$, hence:

Lemma 18. *If σ is an A-unifier for Γ, then the linear diophantine equation $\Lambda(\Gamma)$ has an integer solution $\Lambda(\sigma) = \{x = \Lambda(x\sigma), x \in Dom(\sigma)\}$.*

Proof. Follows from the homomorphism definition. □

A.2 Preprocessing

Now we note that a unification problem $\Gamma = \{u =_A^? v\}$ can not be solved if u and v start with different constant symbols (end with different constant symbols). So we define the following rule

$$\text{Termination} \quad \frac{c_1 u_r =_A^? c_2 v_r \text{ with } c_1 \neq c_2 \text{ and } c_1, c_2 \in \Sigma}{Failure}$$

$$\text{Termination} \quad \frac{u_l c_1 =_A^? v_l c_2 \text{ with } c_1 \neq c_2 \text{ and } c_1, c_2 \in \Sigma}{Failure}$$

Next we can simplify the unification problem by recursively applying the following reduction rule:

$$\text{Cancellation} \quad \frac{[c_1 u =_A^? c_1 v]}{[u =_A^? v]}, \quad \frac{[u c_2 =_A^? v c_2]}{[u =_A^? v]} \quad \text{where } c_1, c_2 \in \Sigma$$

A.3 Enumeration of e-Unifiers

The enumeration of essential unifiers works as in a string unification algorithm for most general unifiers, but all unifiers which contain a part unifier will be eliminated as we go. The important requirement here is that this is decidable.

So the whole process works as follows: Set up the diophantine equation $\Lambda(\Gamma)$ for the given unification problem Γ and for each integer solution compute the corresponding string unifier in the following way. Define for a solution $\alpha : X \mapsto \mathcal{N}$ of a linear diophantine equation $\Lambda(\Gamma)$, a substitution δ_α with $\delta_\alpha(x) = x_1 \ldots x_n$, where $n = \alpha(x)$, $x_i \notin \mathbf{Var}(\sigma)$. Then apply δ_α to u and v and solve

$$\{u\delta_\alpha =^? v\delta_\alpha\}$$

with a standard unification algorithm for terms (such as [57]) rendered appropriately just for strings and the proviso that any variable can only be instantiated by at most one symbol. For the reader familiar with the early unification algorithms for strings, the above integer solution of the translated problem represents the required number of *splittings* of the variables, in the example above with $x = 4$ and $y = 2$ we have $x \mapsto x_1 x_2 x_3 x_4$, $y \mapsto y_1 y_2$. Thus we translate the original A-problem $xby =^? ayayb$ into $x_1 x_2 x_3 x_4 b y_1 y_2 =^? a y_1 y_2 a y_1 y_2 b$ which is then solved such that each variable represents just one symbol as mentioned above.

Let λ be the unifier obtained this way and hence we have the syntactic equation:

$$u\delta_\alpha \lambda = v\delta_\alpha \lambda$$

This enumeration process can be captured in the following generation rule for string unification:

$$\text{GenerationS} \quad \frac{[\Gamma, S], \; \alpha \text{ solves } \Lambda(\Gamma) \text{ and } \lambda \text{ solves } \{u\delta_\alpha =^? v\delta_\alpha\}}{[\Gamma, S \cup \{\delta_\alpha \lambda\}]}$$

where we start with $S = \emptyset$ and recursively apply this rule.

Now let $\Longrightarrow^*$ be the transitive closure of the three rules TCG (*Termination, Cancellation* and *GenerationS*) and correctness and completeness of this standard string unification procedure for mgu's can be shown as follows:

Lemma 19. $[\Gamma, \varnothing] \Longrightarrow^* [\Gamma, S \cup \{\sigma\}]$ iff $\sigma \in \mathcal{U}(\Gamma)$.

Proof. $\Rightarrow$ by definition of *GenerationS*.

$\Leftarrow$ σ is a unifier for Γ implies $\Lambda(\sigma)$ is a solution for $\Lambda(\Gamma)$. Let $\Lambda(\sigma) =: \alpha$. Thus the *GenerationS* is applicable with $\sigma = \delta_\alpha \lambda$ for a λ. $\square$

As a matter of fact the set of unifiers generated by the *GenerationS* is also minimal if an appropriate control is imposed on *GenerationS*. This is also known in the literature on string unification and not of our concern right now. Now we observe a special property of essential string unification, *namely that a part unifier of two strings is always shorter than the unifier it is part of*. Hence as we will see below, the encompassment order for strings is decidable.

Technically we define for the solutions of a linear diophantine equation α, β the ordering $\alpha \leq \beta$ by $\sum_{x \in Dom(\alpha)} \alpha(x) \leq \sum_{x \in Dom(\beta)} \beta(x)$ and show:

Lemma 20. *Let Γ be a string unification problem and λ, σ two unifiers of Γ, then $\lambda \sqsubseteq \sigma$ implies $\Lambda(\lambda) \leq \Lambda(\sigma)$.*

Proof. $\lambda \sqsubseteq \sigma$ implies that there exists μ and ν such that $\sigma = (\mu \lambda \nu)\,|_{\mathbf{Dom}(\sigma)}$.
Thus $\Lambda(\sigma) = \Lambda(\mu) + \Lambda((\lambda \nu)\,|_{\mathbf{Dom}(\lambda)})$, which means that for each $x \in Dom(\sigma)$: $\Lambda(x\sigma) = \Lambda(x\mu) + \Lambda(x(\lambda \nu))$. But then $\Lambda(x(\lambda \nu)) \leq \Lambda(x\sigma)$.
Hence $\Lambda(\lambda) \leq \Lambda(\sigma)$ $\square$

Now we first solve the diophantine equation with some solution α and then we use α to compute the "splitting" substitution δ_α.

Let $uni(\alpha) = \delta_\alpha$ be this function in order to show that every unifier obtained this way is unique:

Lemma 21. *Let $\Gamma = \{u =_A^? v\}$ be an A-unification problem with α a solution for $\Lambda(\Gamma)$, such that there exists $\lambda : X \to \Sigma \cup X$ with a unifier $\delta_\alpha \lambda$ for Γ, then λ is unique.*

Proof. Unification of first order terms is unitary. Hence there exists the function *uni* that maps every solution α of $\Lambda(\Gamma)$ to a unique unifier $uni(\alpha) = \delta_\alpha \lambda$. $\square$

Next we show the length condition for the encompassment order of string unifiers, i.e. the fact that the part unifier is always shorter than the unifier it is part of.

Lemma 22. *Let Γ be an A-unification problem and let $\alpha < \beta$ be two solutions of $\Lambda(\Gamma)$. If there exist two unifiers $uni(\alpha)$ and $uni(\beta)$, then $uni(\beta) \not\sqsubseteq uni(\alpha)$.*

Proof. Suppose by contradiction that $uni(\beta) \sqsubseteq uni(\alpha)$. Note $\Lambda(uni(\beta)) = \beta$ and $\Lambda(uni(\alpha)) = \alpha$. That implies with Lemma 21. $\beta \leq \alpha$, which is a contradiction. $\square$

Finally we observe that the encompassment order for string unifiers is decidable:

Lemma 23. *Let σ,τ be two string unifiers for a given unification problem Γ. Then it is decidable wether $\sigma \sqsubseteq_A \tau$ or $\tau \sqsubseteq_A \sigma$.*

Proof. Let $u \trianglelefteq v$ denote the substring-property, i.e. a string u is a substring of a string v iff $v = v_1 u v_2$. Now by Definition 2 we have $\sigma \sqsubseteq \tau$ iff an instance of σ, e.g. $(\sigma\lambda \mid_{\mathbf{Dom}(\tau)})$ is a sub-substitution of τ, i.e. $\forall x \in \mathbf{Dom}(\sigma) : x\sigma\lambda \trianglelefteq x\tau$. This requirement is the known *substring-matching problem*. More precisely M:= $\{u \trianglelefteq^? v\}$ is a substring matching problem and a solution to M is a substitution λ, such that $u\lambda \trianglelefteq v$.

It can easily be seen that such a λ exists, because $\mid u\lambda \mid \leq \mid v \mid$, hence there are only finitely many attempts to construct λ. The same arguments apply to $\tau \sqsubseteq \sigma$. □

Combining the results of lemmata 19, 20, 21, 22 and 23 we can reformulate the string unification rule *GenerationS* from above into the following rule *Generation*, which can be used to generate all essential string unifiers and only these

$$\frac{[\Gamma, S], \ \alpha \text{ solves } \Lambda(\Gamma) \text{ and } \lambda \text{ solves } \{u\delta_\alpha =^? v\delta_\alpha\} \text{ and } \forall \ \beta \in S : \beta \not\sqsubseteq_A \delta_\alpha\lambda}{[\Gamma, S \cup \{\delta_\alpha\lambda\}]}$$

Now the overall strategy is this: for a given A-unification problem we set up the linear diophantine equations and generate the integer solutions for these equations in some (length) order. Each integer solution determines the number of sub variables (splittings) for each variable in the given unification problem and then we apply some form of a syntactic unification algorithm to this string unification problem. If this gives a unifier σ we check if the set S in the generation rule contains a part unifier of σ: if no we add σ to S, if yes we discard σ and continue.

A controlled algorithm for this kind of enumeration might look like this:

FOR ALL $i \geq 0$ COMPUTE
$S(i) = \{\alpha \mid \alpha \text{ solves } \Lambda(\Gamma), \sum_{x \in Dom(\alpha)} \alpha(x) = i\}$ (* diophantine equations *)
$U(i) = \{uni(\alpha) \mid \alpha \in S(i)\}$ (* real unifiers *)
$E(i) = E \cup \{\lambda \in U(i) \mid \forall \sigma \in E : \sigma \not\sqsubseteq_A \lambda\}$ (* essential unifiers *)
END FOR

---

and we can now state our main result of this paragraph:

Theorem 24. *The above algorithm enumerates all essential unifiers for an A-unification problem Γ.*

Proof. Correctness and Completeness:
follows from Lemma 19.
Essential:
- every generated unifier is unique by Lemma 21.

- A part unifier is always shorter than the unifier it is part of by Lemma 20. Since we generate the set S by length we now just inspect the finitely many unifiers in S thus far generated for the encompassment relation.

- this is decidable by Lemma 20. □

A.4 A note on termination if the set of essential unifiers is finite

A well known problem with most naive string unification algorithms is that they may not terminate even if the set of unifiers for the given unification problem happens to be finite and this is unfortunately also true for our enumeration process for essentials. But see [36] and others for more recent works with elaborate termination criteria built upon Makanin's algorithm, or on Plandowski's algorithm [23]. These are important and interesting indeed, but unfortunately all of these - including our own results in this paper - are not very helpful for a practical reasoning system yet.

The termination criterion for the enumeration algorithm above is as follows:

For an equational problem Γ modulo A and a finite set of unifiers $S \subset U\Sigma(\Gamma)$ is the predicate $\mathcal{P}(S,\Gamma) := \exists \sigma \in \mathcal{U}\Sigma(\Gamma) : \forall \beta \in S : \beta \not\sqsubseteq \sigma$ decidable?

Within the enumeration algorithm above this can be expressed more directly in the following way:

after the update of the "collector set" $E(i) = \{\sigma_1, \sigma_2, ..., \sigma_k\}$ decide with $n_j = |\mathbf{Dom}(\sigma_j)|, j \leq k$ the solvability of the system:

$$\Xi = \begin{cases} u =^? v \\ \{x_{11}\sigma_1 \not\sqsubseteq x_{11}\} \vee \{x_{12}\sigma_1 \not\sqsubseteq x_{12}\} \vee .... \vee \{x_{1n_1}\sigma_1 \not\sqsubseteq x_{1n_1}\} \\ .... \\ \{x_{k1}\sigma_k \not\sqsubseteq x_{k1}\} \vee \{x_{k2}\sigma_k \not\sqsubseteq x_{k2}\} \vee .... \vee \{x_{kn_k}\sigma_k \not\sqsubseteq x_{kn_k}\} \end{cases}$$

If Ξ has no solution, then there is no unifier for Γ, which does **not** encompass any of the essential unifiers in $E(i)$ and the enumeration stops, otherwise the process continues. It is known that the substring relationship is not expressible as a boolean formula of string equations and -inequations (see e.g. [35]). As a consequence the encompassment relation is also not expressible and hence it is not known, wether this termination criterion can be built into a decidability method, such as in Makanin's algorithm.

Acknowledgements

We like to thank our first reviewer at the unification workshop in 2008 and the interesting discussion there and afterwards with several participants of the workshop. All of this led to a complete reformulation of our basic definitions and greatly simplified the proofs and the general presentation, finally leading to our more general framework based on the encompassment order as presented here and in [67]. We also acknowledge the very critical and competent later reviews of this paper. We are also indebted to Artur Jez´ substantial contribution to paragraph 3.1, where he pointed to a serious flaw in our first version of this paper (the unitary, finitary result).

References

[1] RTA list of open problems. http://rtaloop.pps.jussieu.fr, 2008.

[2] S. I. Adian and V. G. Durnev. Decision problems for groups and semigroups. *Russian Mathematical Surveys 55.2 (2000): 207.*, 55 (2):207, 2000.

[3] F. Baader. Unification in idempotent semigroups is of type zero. *Journal of Automated Reasoning*, 2(3):283–286, 1986.

[4] F. Baader. A note on unification type zero. *Information Processing Letters*, 27:91–93, 1988.

[5] F. Baader and T. Nipkow. *Term Rewriting and all That*. Cambridge University Press, 1998.

[6] F. Baader and J. Siekmann. General unification theory. In D. Gabbay, C. Hogger, and J. Robinson, editors, *Handbook of Logic in Artificial Intelligence and Logic Programming*, pages 41–126. Oxford University Press, 1994.

[7] F. Baader and W. Snyder. Unification theory. In A. Robinson and A. Voronkov, editors, *Handbook of automated reasoning, vol 1*. Elsevier Science Publishers, 2001.

[8] H. J. Buerckert, A. Herold, D. Kapur, J. Siekmann, M. Stickel, M. Tepp, and H. Zhang. Opening the AC-unification race. *Journal of Automated Reasoning 4.4 (1988): 465-474.*, 4 (4):465–474, 1988.

[9] H.-J. Buerckert, A. Herold, and M. Schmidt-Schauss. On equational theories, unification and (un)decidability. *Journal of Symbolic Computation*, 8:3–49, 1989.

[10] V.K. Bulitko. Equations and inequalities in a free group and semigroup. *Geometr. i Algebra Tul. Gos. Ped. Inst. Ucen. Zap. Mat. Kafedr.*, 2:242 – 252, 1970.

[11] W. Charatonik and L. Pacholski. Word equations with two variables. In *Word Equations and Related Topics, IWWERT*, pages 43–56. Springer Berlin Heidelberg, 1993.

[12] Ch. Choffrut and J. Karhumaeki. Combinatorics of words. In *Handbook of formal languages*, pages 329–438. Springer Berlin Heidelberg, 1997.

[13] R. Dabrowski and W. Plandowski. On word equations in one variable. *Lecture Notes in Computer Science*, 2420:212–220 and in: Algorithmica, vol 60, no 4, p. 819–828, 2002 and 2011.

[14] N. Dershowitz. Termination of rewriting. *Journal of symbolic computation*, 3 (1):69–115, 1987.

[15] N. Dershowitz and J.-P. Jouannaud. Rewrite systems. In J. van Leeuwen, editor, *Handbook of Theoretical Computer Science*, pages 244–320. Elsevier Science Publishers (North-Holland), 1990.

[16] V. Diekert. Makanins algorithm. In M. Lothaire, editor, *Algebraic Combinatorics on Words*, chapter 12, pages 387–442. Cambridge University Press, 2002.

[17] V. Diekert, A. Jez, and W. Plandowski. Finding all solutions of equations in free groups and monoids with involution. In *Computer Science-Theory and Applications: 9th International Computer Science Symposium in Russia, (CSR)*, pages 1 – 15. Springer LNCS vol 8476, 2014.

[18] E. Domenjoud. A technical note on AC-unification: The number of minimal unifiers of the AC equation. *Journal of Automated Reasoning*, 8 (1):39–44, 1992.

[19] D. J. Dougherty and P. Johann. An improved general E-unification method. *J of Symbolic Computation*, 11:1–19, 1994.

[20] V. Durnev. Studying algorithmic problems for free semi-groups and groups. In *Logical Foundations of Computer Science*, pages 88–101. Springer Berlin Heidelberg, 1997, 1997.

[21] V. G. Durnev. On equations in free semigroups and groups. *Mathematical Notes of the Academy of Sciences of the USSR*, 16 (5):1024–1028, 1974.

[22] E. Eder. Properties of substitutions and unifications. *Journal of Symbolic Computatio*, 1(1):31–46, 1985.

[23] F. Fages and G. Huet. Complete sets of unifiers and matchers in equational theories. *Theoretical Computer Science*, 43(1):189–200, 1986.

[24] Jean H. Gallier. Unification procedures in automated deduction methods based on matings: a survey. Technical Report CIS-436, University of Pensylvania, Dep of Computer and Information Science, 1991.

[25] C. Gutierrez. Solving equations in strings: On Makanin's algorithm. *LATIN'98: Theoretical Informatics. Springer Berlin Heidelberg*, pages 358–373, 1998.

[26] A. Herold and J. Siekmann. Unification in Abelian semigroups. *Journal of Automated Reasoning*, 3:247–283, 1987.

[27] J. I. Hmelevskij. The solution of certain systems of word equations. *Dokl. Akad. Nauk SSSR Soviet Math.*, 5:724 pp, 1964.

[28] J. I. Hmelevskij. Word equations without coefficients. *Soviet Math. Dokl.*, 7:1611–1613,

1966.

[29] J. I. Hmelevskij. Solution of word equations in three unknowns. *Dokl. Akad. Nauk SSSR Soviet Math.*, 5:177 pp, 1967.

[30] M. Hoche, J. Siekmann, and P. Szabo. String unification is essentially infinitary. In Mircea Marin, editor, *The 22nd International Workshop on Unification (UNIF'08)*, pages 82–102, Hagenberg, Austria, 2008.

[31] M. Hoche and P. Szabo. Essential unifiers. *Journal of Applied Logic*, 4(1):1–25, 2006.

[32] J. M. Howie. *An Introduction to Semigroup Theory.* Academic Press, 1976.

[33] G. Huet. A complete proof of correctness of the knuth-bendix completion algorithm. *Journal of Computer and System Sciences*, 23 (1):11–21, 1981.

[34] G. Huet. Higher order unification 30 years later. In *Theorem Proving in Higher Order Logics.*, volume 2410, pages 3–12. Springer LNCS, 2002.

[35] L. Ilie. Sub-words and power-free words are not expressible by word equations. *Fundamenta Informaticae*, 38:109–118, 1999.

[36] J Jaffar. Minimal and complete word unification. *JACM*, 27 (1):47–85, 1990.

[37] A. Jez. One-variable word equation in linear time. *ICALP, Lecture Notes in Computer Science*, vol 7966:324–335, 2013.

[38] A. Jez. Recompression: a simple and powerful technique for word equations. In N. Portier and T. Wilke, editors, *STACS*, volume 20 of *LIPIcs*, pages 233–214 and arXiv: 1203.3705v3. Schloss Dagstuhl Leibniz Zentrum for Informatics, 2013.

[39] D. Kapur and P. Narendran. Complexity of unification problems with associative-commutative operators. *Journal of Automated Reasoning*, 9 (2):261–288, 1992.

[40] D. Kapur and P. Narendran. Double-exponential complexity of computing a complete set of AC-unifiers. In *Proceedings of the Seventh Annual IEEE Symposium on Logic in Computer Science, LICS92*, 1992.

[41] C. Kirchner and H. Kirchner. Rewriting solving proving. http://www.loria.fr/ ckirchne/=rsp/rsp.pdf, 2006.

[42] K. Knight. Unification: A multidisciplinary survey. *ACM Computing Surveys (CSUR)*, 21 (1):93–124, 1989.

[43] A. Koscielski and L. Pacholski. Complexity of Makanin's algorithm. *JACM*, 43 (4):670–684, 1996.

[44] M. Laine and W. Plandowski. Word equations with one unknown. *International Journal of Foundations of Computer Science*, 122 (2):345–375, 2011.

[45] A. Lentin. Equations in free monoids. In *M. Nivat; Automata, Languages and Programming*. North Holland, 1972.

[46] A. Lentin and M.P. Schuetzenberger. A combinatorial problem in the theory of free monoids. In R.C. Bose and T.E. Dowling, editors, *Combinatorial mathematics*, pages 112–144. Univ of North Carolina Press, 1967.

[47] M. Livesey and J. Siekmann. Termination and decidability results for string unification. report Memo CSM-12, Computer Centre, Essex University, England, 1975.

[48] M. Lothaire. *Combinatorics on words*, volume 17 of *Encyclopedia of Mathematics*. Addison-Wesley1997, reprinted in: Cambridge University Press, Cambridge mathematical library, 1983.

[49] M. Lothaire. *Algebraic combinatorics on words*. Cambridge University Press, 2002.

[50] G.S. Makanin. The problem of solvability of equations in a free semigroup. *Original in Russian: Mathematicheskii Sbornik, 103(2), 1977, pp. 147-236; Math. USSR Sbornik*, 32:129–198, 1977.

[51] A. A. Markov. Theory of algorithms. *Trudy Mathematicheskogo Instituta Imeni VA Steklova, Izdat.Akad. Nauk SSSR*, 17:1038, 1954.

[52] Y. Matiyasevich. The connection between Hilbert's 10th problem and systems of equations between words and lengths. *Seminars in Mathematics*, 8:61–67, 1970.

[53] E. Obono, S. Goralcik, and P. M. Maksimenko. Efficient solving of the word equations in one variable. In *Proceedings of MFCS'94*, pages 336–341, 1994.

[54] W. Plandowski. Satisfiability of word equations with constants is in Pspace. *JACM*, 51, 3:483–496, 2004.

[55] G. Plotkin. Building-in equational theories. In B. Meltzer and D. Michie, editors, *Machine Intelligence*, volume 7, pages 73–90. Edinburgh University Press, 1972.

[56] P. Raulefs, J. Siekmann, P. Szabo, and E. Unvericht. A short survey on the state of the art in matching and unification problems. *ACM Sigsam Bulletin*, 13 (2):14–20, 1979.

[57] J. A. Robinson. A machine-oriented logic based on the resolution principle. *Journal of the ACM*, 12(1):23–41, 1965.

[58] J. A. Robinson. Computational logic: the unification computation. In B. Meltzer and D. Michie, editors, *Machine Intelligence*, volume 6, pages 63–72. Edinburgh University Press, 1970.

[59] Manfred Schmidt-Schauss. Unification under associativity and idempotence is of type nullary. *Journal of Automated Reasoning*, 2(3):277–281, 1986.

[60] K. Schulz. Word unification and transformations of generalized equations. *J. of Autom. Reasoning*, 11:149–184, 1993.

[61] K. U. Schulz. Makanin's algorithm for word equations: Two improvements and a generalization. In *Proceedings of the First International Workshop on Word Equations and Related Topics (IWWERT90)*, volume 572, pages 85–150. Springer LNCS, 1992.

[62] J. Siekmann. String unification. report CSM-7, Computer Centre, University of Essex, 1975.

[63] J. Siekmann. *Unification and matching problems*. PhD thesis, Essex University, Computer Science, 1976.

[64] J. Siekmann. Universal unification. In *Proceedings of the 7th International Conference on Automated Deduction*, pages 1–42. Springer London, 1984.

[65] J. Siekmann. Unification theory. *Journal of Symbolic Computation*, 7(3 & 4):207–274, 1989.

[66] Mark E. Stickel. A unification algorithm for associative-commutative functions. *Journal of the ACM (JACM)*, 28.3:423–434, 1981.

[67] P. Szabo, J. Siekmann, and M. Hoche. What is essential unification? In *Martin Davis on Computability, Computation, and Computational Logic.* Springer's Series "Outstanding Contributions to Logic", 2016.

[68] P. Szabo and E. Unvericht. D-unification has infinitely many mgus. Technical report, University of Karlsruhe, Inst. f. Informatik I, 1982.

[69] Peter Szabo. *Unifikationstheorie erster Ordnung.* PhD thesis, University Karlsruhe, 1982.

Quantum States and Disjunctive Attacks in Talmudic Logic

Michael Abraham, Israel Belfer, Uri Schild
Bar-Ilan University, Israel

Dov Gabbay
Ashkelon Academic College, Israel
Bar-Ilan University, Israel
King's College London
University of Luxembourg

This paper provides logical modelling for the results contained in the twelfth monograph on Talmudic logic entitled *Fuzzy Logic and Quantum States in Talmudic Reasoning* [2].[1]

This paper directly impacts on abstract argumentation theory, temporal and fuzzy arguments and disjunctive collapse. It deals with attacks on a target set of arguments which results in the target to be considered in a quantum like superposition state. The attack is not crisp enough and so cannot be said to be focussed on any individual member or any clear subset of the target. As a result the target set needs to be treated like a quantum superposition of its members.

1 Background and orientation

We begin our discussion with several examples.

We thank the referees for most valuable comments

[1] As we have indicated in our first paper and in our book [1] on Talmudic Logic, the aim of this series (of possibly 25-30 books) is twofold:

1. Import logical tools to the service of modelling and explaining Talmudic reasoning and debate.
2. Export ideas and logical constructions from Talmudic debate for the application and use in general logical theory, artificial intelligence and agency and norms.

Example 1.1 (Disjunctive attacks: Story 1). *Mr. Smith is a rich old man who wants to donate a very rare classic painting to one of two national museums. He committed the donation in a letter to the two museums and copied and approved by Charity Commission, so the donation to one of the two museums was legally done, accepted and in force, except that the choice as to which of the two museums the painting will be given has not been made yet. Mr. Smith said that he would inform the Charity Commission and the museums which museum he would choose in a few days. The donation is in force, however, regardless the status of the choice. Mr. Smith unfortunately died before he made that choice. We are now left with an unclear legal situation regarding ownership. Let a, b and x denoted as follows:*

$b =$ *the painting does belong to museum b*
$a =$ *the painting does belong to museum a*
$x =$ *body of laws regarding ownership.*

We have, of course, that a and b are mutually exclusive. Therefore we have that x disjunctively attacks (see [6]) the set $\{a, b\}$. The attack says one of $\{a, b\}$ must be false. We must be clear here.

Suppose we are dealing with n museums. The options are then

$$a_i = \text{the painting belongs to the } i\text{-th museum}, i = 1, ...n.$$

Then we have that x implies that exactly on of a_i holds.

Put differently, x implies the set $\{a_i | i = 1, .., n\}$, where the meaning of imply a set of formulas is that exactly one of a_i is true, or equivalently that exactly one $\neg a_i$ is false.

Then again reformulating we can say that x attacks the set $\{\neg a_i | i = 1, ..., n\}$, where the meaning of attacking a set is that exactly one member $\neg a_i$ is false, namely exactly one a_i is true.

In case $n = 2$, we have $a = \neg b$ and $b = \neg a$ and so we have that x attacks $\{\neg a, \neg b\}$ is the same as x attacks $\{b, a\} = \{a, b\}$. Talmudic logic debate distinguishes several views on this scenario. The facts on the ground are that the museum's claim that there was a legally binding donation and as for the question of who is beneficiary, a or b, a reasonable deal can be worked out, such as an agreed arrangement of co-ownership, or sharing, or we can let the estate of Mr. Smith continue and choose a museum or we can flip a coin, or ...whatever other symmetrically reasonable solution.

Talmudic logic debate offers two main views on this:

View 1. Quantum like view. *This view is that, since Mr. Smith died before making a choice of a museum, ownership is superimposed evenly on both museums,* in the same sense as, nowadays, modern quantum mechanics treats the two slits experiment [5]. Recall that in the two slit experiment a single electron is sent towards two slits a and b and the electron passes through both slits as a wave and interferes with itself. So even though logically in classical mechanics the electron is expected to pass only through one slit, it is also a wave according to quantum mechanics and so it passes through both.

The Talmudic debaters holding this view are divided in their verdict:

Option 1. Since a and b are mutually exclusive, there is no longer a donation. The superposition of ownership cancels the donation. The actual Talmudic debate is in connection with marriages but we have adapted the story to Mr Smith and his donation of paintings. See Talmud Bavli, Kidushin, Page 51a and after.

Option 2. The superposition of ownership does not cancel the donation. There is a donation the superposition holds but the fact is that the superposition causes lack of clarity of what to do and it should be undone by court order. The museums should waive their "ownership" back to the estate for otherwise the normal flow of life would be disrupted. After that, the estate can re-donate the painting if they want to.

Of course, those options have implications towards estate tax duties, etc.

Note that both options agree that ownership is super-imposed on both $\{a,b\}$. They differ in their verdict.

View 2. Fuzzy probabilistic view. *There is no superposition. There was a donation and we view the scenario as if there was a choice of a museum, except that we do not know what it was, i.e. we treat the case as if Mr. Smith did choose a museum, wrote a letter but died and the letter was lost). So we have a case of purely epistemic uncertainty here and we are expected to provide some mechanism to divide/allocate the painting. For example:*

1. *Share ownership 50/50.*

2. *Make a case for one museum over the other, for example, if the painting was in the special area of museum a, then we can argue and reasonably claim that there is high probability that a was chosen.*

3. *Recommend other arrangements, such as decision by lottery, or time sharing, etc.*

Note that there are further implications to View 2. For example if the number of museums involved is very large, we could on probabilistic grounds, agree that the painting remains with the estate, as the probability for each museum to be the owner is very low. This is an interesting Talmudic view. We might think that it is possible for all the museums to form a coalition and ask for the painting. The Talmud will not allow this. To explain this aspect of this view, think of a different scenarioes.

Scenario 2.1:(Compare with Example 1.2). *There is one painting which was donated to one museum, from among many paintings and we do not know which one it is. Say both the donator and the museum curator die suddenly. The problem is whether we forbid the estate owners of these paintings to sell any of their paintings for fear that it is the one belonging/donated to the museum. The Talmud view in this case is that since the majority of paintings was not donated we allow the sale.*

Compare this scenario with the following variation:

Scenario 2.2. *This scenario is the case of donating one painting but not yet deciding which one, and before a decision is made, the owner dies. In this case, the Talmud says that each of the paintings could have been chosen, and so the museum is part owner/potential owner in each painting and so none of them can be sold! This is like modern quantum superposition view.*

There are other contexts where this practical probabilistic reasoning makes sense. If one Ebola infected person passed through an airport around the time when there were 2000 others present, we can assume about each of the others that he/she is not infected but cannot treat them as such.

We now conclude our discussion of View 1 and View 2 of the story of Mr. Smith donation of one painting to one of two museums. The main thrust of the story is that there is an attack on the set $\{a, b\}$ without there being any specific attacks on a or on b. The story can continue as follows:

Suppose each of a and of b, independently attacks c, the details of the attack are not important (maybe c is an art critic claiming the painting is a forgery). What is important are the formal options for handling the situation. We have several options for reasoning here

1. *c must be out (i.e. false), since either a or b is in (i.e. true) and both attack c.*

2. *c must be in, since the disjunctive attack is super-imposed on both a and b, so neither is safely to be considered in (true).*

3. *c is undecided since we do not know exactly what is going on with $\{a, b\}$.*

4. c joins $\{a,b\}$ in the status of being a member of the superposition set. In other words, we have that if x disjunctively attacks $\{a,b\}$ and a attacks c, and the attack of x on $\{a,b\}$ is perceived as a superposition on $\{a,b\}$, then the constellation of [x disjunctively attacks $\{a,b\}$ and a attacks c, and the attack of x on $\{a,b\}$ is perceived as a superposition on $\{a,b\}$] is taken to be equivalent to the constellation [x attacks $\{a,b,c\}$ and the attack is perceived as a superposition of $\{a,b,c\}$].

Example 1.2 (Disjunctive attack Story 2). *Mr. Smith is a rich man owning 2 original masterpieces. He decides to donate one of these paintings to the museum (a charity). There are several steps to be taken to accomplish this properly. Select the painting, transfer ownership, put conditions on its use and exhibition, get tax relief on the donation, etc., etc.*

These steps are persistent in time. Once accomplished they remain so. So the temporal flow is to execute each step properly and then legally end up with the result. The Talmudic scenario is to study, debate and rule in cases where the steps become fuzzy. The question is then to determine what final result we have in this case. The logic behind the Talmudic debate of the various scenarios is the Talmudic fuzzy logic and Talmudic disjunctive attacks.

Scenario 3. *Mr. Smith commits a painting to the museum. The museum sends Mr. Jones to go with Mr. Smith to the "storage vault" and choose a painting.*

Storyline 1. *On the way both die (tragic traffic accident).*

Question 1. *What does the museum get/own? What would the heirs/estate of Mr. Smith do?*

Storyline 2. *Mr. Smith and Mr. Jones get to the storage and choose a painting. On the say out of the storage they both die. So we know a painting was chosen but we do not know which one and we have no way of knowing.*

Question. *Same as before.*

Storyline 3. *Mr. Smith authorises Mr. Jones to go to the storage and choose a painting. Mr. Jones does that and telephones Mr. Smith and tells him what he chose. A few minutes later Mr. Smith dies of a heart attack and Mr. Jones dies in a tragic accident. The museum knows the government was secretly and unlawfully* recording all telephone conversations of prominent citizens. They could try and get the recording of which painting was chosen. This is very difficult because the Government will never admit that it is listening to its citizens. In this scenario, we could find out what painting was chosen, but for all practical purposes, we find ourselves in Storyline 2.

We note that once we are in a state of superposition, like when a painting was donated to one of two museums but not decided which one or one of two paintings was donated to a single museum but not decided which one, we can collapse the superposition retrospectively by, for example, flipping a coin. This is parallel to quantum superposition which can collapse when we do measurements.

Let us now analyse these stories. Let

$$S = \{\pi, \pi'\}$$

be the set of paintings and let $G(x)$ be the predicate that x was given by Mr. Smith to the museum. First we ask: do we know for sure that $\exists x G(x)$ must hold? The problem is that if no painting was chosen, was there a donation?

If we decide that there was a donation, then which painting? Can the museum sell something? Can the museum transfer to another legal entity whatever it has?

Storyline 1. *This is the case where a painting was donated but none was chosen. Compare with Example 1.1.*

Rava opinion. *There is no deal. The museum gets nothing. (Compare with View 1 of Example 1.1.)*

Abeyei opinion. *There was a valid deal. $\exists x G(x)$ is true but we are in doubt as to which painting was given to the museum. We have a case of superposition here.*
According to this view, we can recommend some options.

A1: The museum is to give up voluntarily the donation. This is what the law forces them to do.

A2: Alternatively, in practice, they may reach some deal.

  1. The estate of Mr. Smith can donate all the paintings to the museum.
  2. Choose a paining now.
  3. Rotate the donation, rotate every season a different painting.
  4. etc.

This may be OK for paintings and museums, but there are other scenarios which are less flexible. Mr. Smith may have two beautiful daughters and he has agreed to give one of his daughters in marriage to Mr. Jones' son. According to Abayei's approach, only option A1 can be taken. No sharing or rotation or anything is possible, only divorce from each of them. According to the law one cannot be married to two sisters at the same time. One cannot even choose one later, because the new choice may not be the correct one and if married to one you cannot have a marital relationship with her sister. If we look at Storyline 2, here there was a choice of painting or daughter, but we do not know which one. So we can apply a different logical machinery to this case. Maybe we can argue that the museum has all the paintings of Van Gogh except the one which Mr. Smith owns and so it is most likely that the last van Gogh was chosen or in the case of marriage, one can argue that perhaps one of the daughters already knows Mr. Jones' son and the process was most likely aimed at choosing her?

To sharpen the difference between Storyline 1 Abayei and Storyline 2 Abayei, we note the following:

Our storyline 1, Abayei, we accept that $\exists x G(x)$ holds but we do not accept for any $x \in S$ that $G(x)$ holds in a clear cut way, as opposed to some fuzzy way. So Rava says there is no engagement and Abayei says that there is, but it is fuzzy. In Storyline 2, we also accept that for one of $x \in S, G(x)$ holds, but we do not know which one.

We need a logic which can model such distinctions!

2 Argumentation networks

We need to model the above examples. We shall use a version of disjunctive argumentation networks [6, 3].

Definition 2.1. *A finite argumentation network has the form (S, R), where S is a finite non-empty set of arguments, and $R \subseteq S \times S$ is an attack relation. We also write $x \twoheadrightarrow y$ in diagrams to express xRy, x attacks y.*

Example 2.2. *Imagine two pairs of parents planning a joint wedding for their children. They need to compose a list of guests of several types.*

1. *Relatives from each family*

2. *Neighbours and friends of parents*

3. *Friends of the bride and bridegroom*

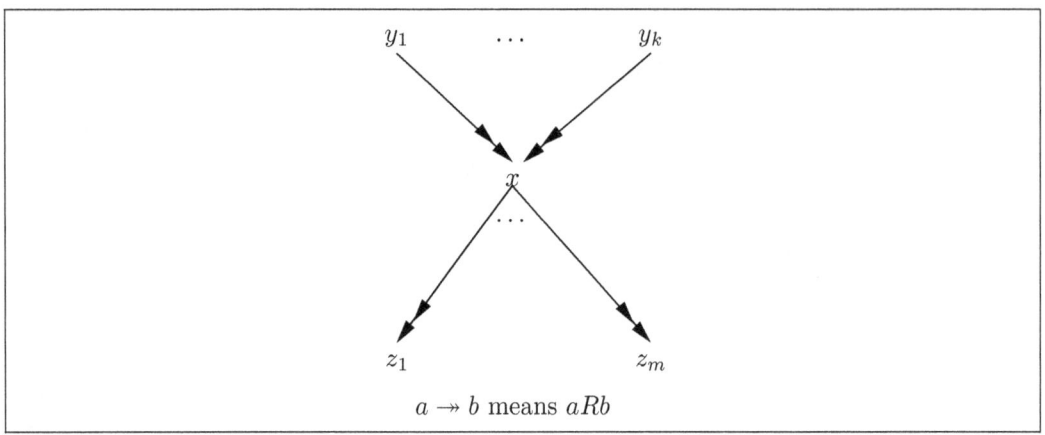

Figure 1

4. Colleagues and co-workers

Inviting family can be a problem!

Auntie Bertha might say "I am not coming if that bastard ex-husband of mine is invited". I.e., Bertha $\twoheadrightarrow$ ex-husband.

Grandma Teresa might say "I don't want these kids inviting too many of these hippy crazy friends of theirs, espeically not the drummers". I.e., Teresa $\twoheadrightarrow$ {set of hippies}.

Figure 1 can describe the problematic map which exists:

x is one possible invitee, say Grandma Teresa. She is 109 years old and $y_1, \ldots, y_k$ object to inviting her. Possibly because she is too old and they are worried about her health The reason does not matter. The important fact here is the double arrow $y_i \twoheadrightarrow x$. This means y_i wants x out. So if y_i is invited, x cannot be invited. Similarly, x objects to $z_1, \ldots, z_m$. So the Figure 1 describes the entire configuration around x. We want to define a maximal set E of invited guests such that the following holds:

1. *$x, y \in E \Rightarrow x$ does not attack y, (i.e., not xRy). I.e., E is conflict free. No member x of E says "I object" to another member of E.*

2. *If any x says "why did you invite $z \in$ and you did not invite me? How could you invite this terrible person z"? (i.e., we have $z \twoheadrightarrow z$), then we can say, "we had to invite $y \in E$ and unfortunately, y was against you x" (i.e. for some $y \in E, y \twoheadrightarrow x$).*

Such a set E which is also maximal, is called in the argumentation community "a preferred extension". These always exist.

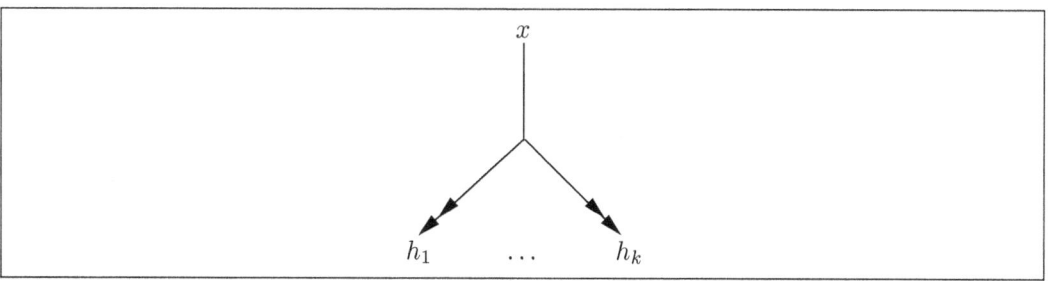

Figure 2

A disjunctive attack has the form $x \twoheadrightarrow H$ where $H \subseteq S$. Its meaning is

- if $x \in E$ then for some $y \in H (y \notin E)$.

This means if you invite x then one of H must not be invited. For example x may be having an affair with both (h_1, h_2). So it is bad taste to invite both. We know about it, but h_1 and h_2 do not know about each other, so it is better not to have them both, says x. We use the notation of Figure 2

Definition 2.3 (See [6], Definition 3.3).

1. *A finite disjunctive argumentation network has the form $\mathcal{A} = (S, \rho)$, where S is a finite set of arguments and $\rho \subseteq S \times (2^S - \emptyset)$, i.e. ρ is a relation of (disjunctive attacks) between elements $x \in S$ and non-empty subsets $H \subseteq S$ denoted as $(x\rho H)$.*

   Let (S, ρ) be a network and let $E \subseteq S$:

   (a) *We say E is conflict free iff for no $x \in E$ and $H \subseteq E$ do we have $x\rho H$.*

   (b) *We say that E protects α iff for any $z\rho H \cup \{\alpha\}$ there exists a $\beta \in E$ and $E_3 \subseteq E$ and $H_3 \subseteq H$ such that $\beta \rho H \cup H_3 \cup E_3 \cup \{z\}$.*

   (c) *We say E protects itself if it protects each of its members.*

   (d) *We say E is a complete extension if E is conflict free, protects itself and contains all the elements it protects.*

Talmudic attack $x\rho H$ wants exactly one $y \in H$ to be out. Talmudic logic thinks of it as a collapse of $x\rho H$ to xRy.

The next definition, 3.1 will explain what we mean by collapse, and give a more correct way to obtain the complete extensions according to Talmudic logic.

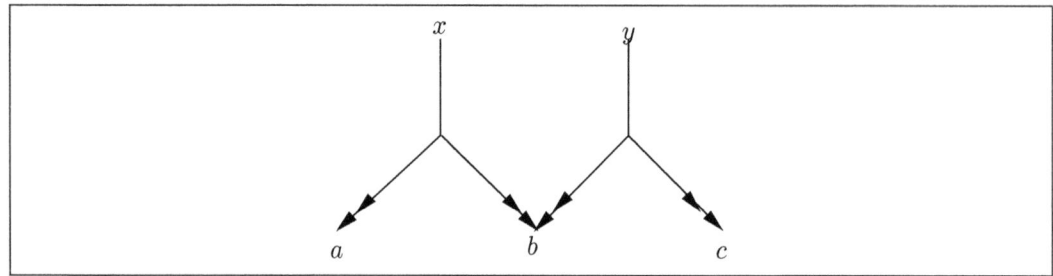

Figure 3

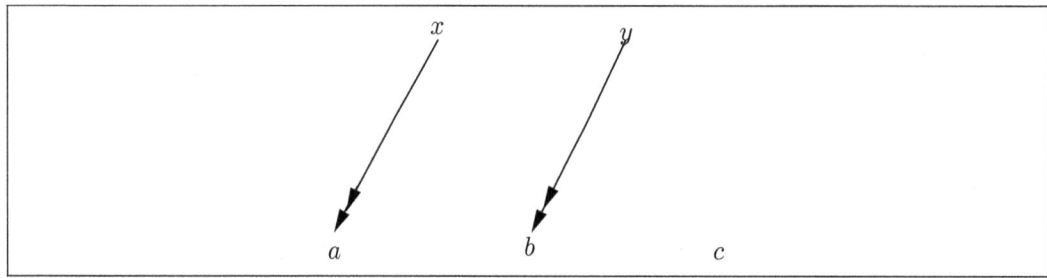

Figure 4

3 Talmudic argumentation systems

Definition 3.1. *Let $\mathcal{A}$ be a finite disjunctive network and let $x\rho H$ be one of its attacks. We say that a set $\mathbb{F}((x,H))$ is a collapse set for (x,H) if it is the set of all $\mathcal{A}_y, y \in H$ of the form $\mathcal{A}_y = (S, \rho_y^x)$, where $\rho_y^x = (\rho - \{x, H)\}) \cup \{(x, \{y\})\}$. In other words, (S, ρ_y^x) is the network where $x\rho H$ is replaced by $x\rho\{y\}$, i.e. $x\rho H$ collapses to $x\rho\{y\}$.*

For each $x\rho H$, let $\mathbf{f}(x, H)$ choose one pair $(x, y), y \in H$. Let $\mathcal{A}_\mathbf{f}$ be the total collapse of $\mathcal{A}$ according to $\mathbf{f}$, defined as $(S, R_\mathbf{f})$, where $R_\mathbf{f} = \{\mathbf{f}(x, H) | x\rho H\}$.

Example 3.2.

1. **Complete collapse.** *Consider the network of Figure 3.*

   Here we have $x\rho\{a, b\}$ and $y\rho\{b, c\}$. The total collapses are the networks in Figures 4, 5, 6 and 7.

2. **Partial collapse.** *We may have that say $x\rho\{a, b\}$ collapses while $y\rho\{b, c\}$ does not collapse. So we have in this case the possible Figures 8 and 9.*

Remark 3.3. *We have to decide what the Talmud would say about attacks emanating from non-collapesed nodes. Consider Figure 10*

Disjunctive Attacks in Talmudic Logic

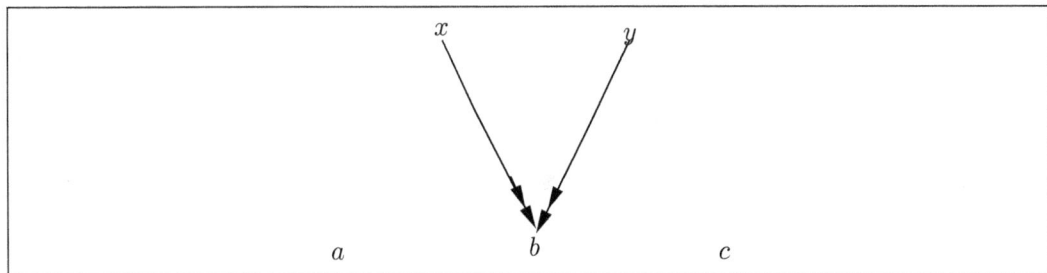

Figure 5

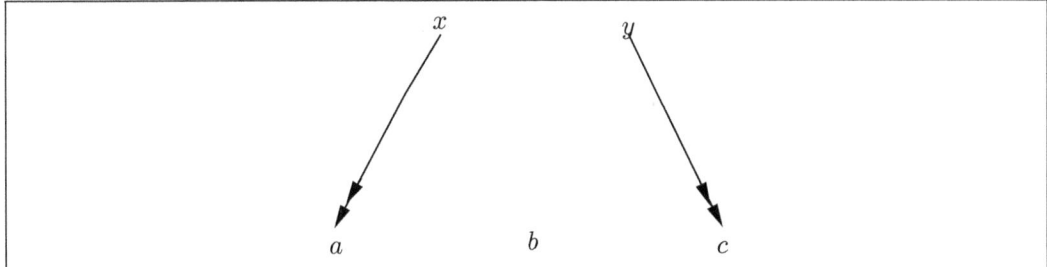

Figure 6

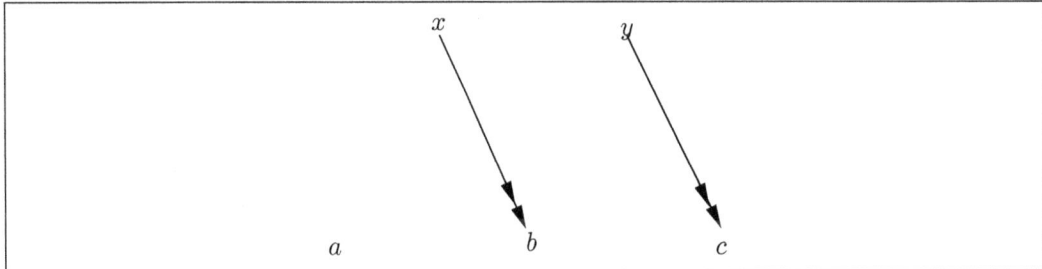

Figure 7

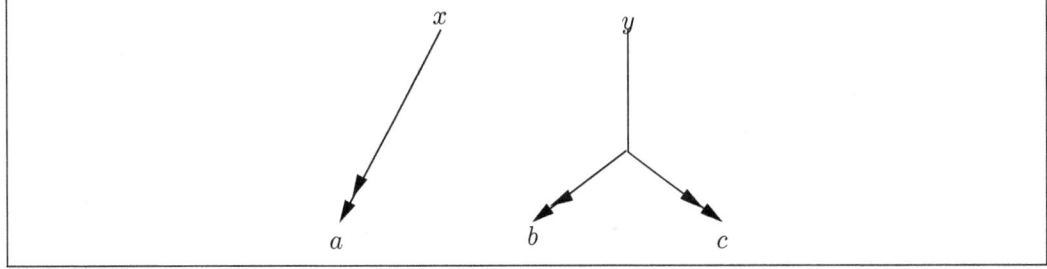

Figure 8

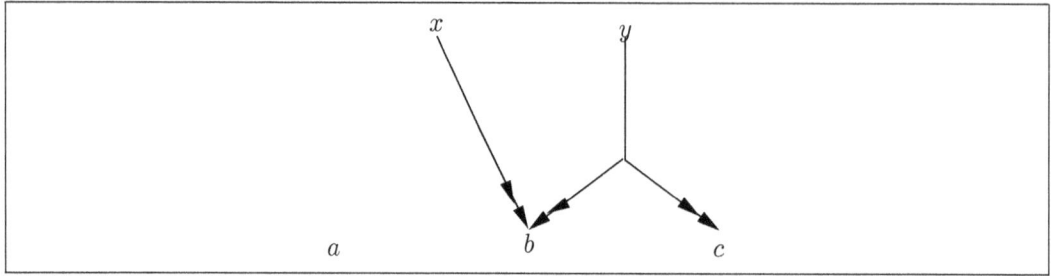

Figure 9

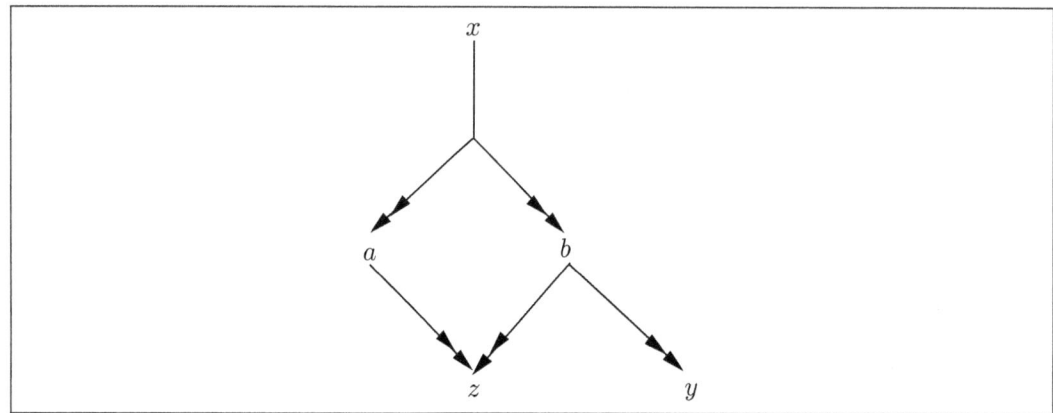

Figure 10

In this figure the attack of x on $\{a,b\}$ remains uncollapsed. So this is the final fixed figure. What is our view of $\{a,b\}$? Do we consider them as both in/true (since there is no collapse) for the purpose of the attacks $b \twoheadrightarrow y$, $b \twoheadrightarrow z$ and $a \twoheadrightarrow z$? Or do we regard then as undecided? Do we give them fuzzy values?

The Talmud approach can be modelled by four values $\{in, out, undecided, wave\}$. So we use labelling $x \in \{in, out, und, wave\}$.

So, in Figure 10 we may have that $\{a,b\}$ does not collapse, so we give a, b value "wave" each. This value is passed on to y and z.

If y or z further attack some nodes, they will pass on the value "wave" to their targets.

Remark 3.4. *Compare with the traditional Caminada labellings and other approaches in [4]. Let us look again at Figure 1 where $y_1, \ldots, y_k$ are all the attackers of x and let us write the conditions on any $\lambda : S \mapsto \{in, out, und, wave\}$ to be a legitimate Talmudic labelling for a traditional network (S, R) without disjunctive attacks.*

(TC1) $\lambda(x) = out$, if for some $y_i, \lambda(y_i) = in$.

(TC2) $\lambda(x) = in$, if for all $y_i, \lambda(y_i) = out$.

(TC3) $\lambda(x) = und$, if none of y_i has value $\lambda(y_i) = in$ and some of $\lambda(y_i) = und$.

(TC4) $\lambda(x) = wave$, if none of $\lambda(y_i) = in$ and none of $\lambda(y_i)$ is und and some of $\lambda(y_i) = wave$.

Remark 3.5. *We now have to define what is a legitimate λ for a network (S, ρ) with disjunctive attacks $\rho \subseteq S \times (2^S - \varnothing)$. We shall reduce this concept by induction to the traditional case with four values as defined in Remark 3.4. The reduction is by induction on the number of disjunctive attacks in (S, ρ). We first need a concept of constraints on λ.*

1. *Let (S, R) be an argumentation network of any kind (traditional or Talmudic) with $R \subseteq S \times S$. Let λ_1 be a partial function $\lambda_1 :$ Subset E of $S \mapsto$ values. We say λ is a legitimate extension under the constraint λ_1 if λ is legitimate and λ agrees with λ_1 on its values.*

2. *For example in the configuration of Figure 1 we may have the constraint $\lambda_1(y_1) = wave$. However, if the figure is part of a larger network and y_1 is attacked by a node which needs to be in, then λ cannot overrule λ_1 on the value of y_1.*

   When we have a constraint λ_1 it may be the case that no legitimate λ exists with such a constraint.

3. *We now define what it means to be a legitimate Talmudic extension for (S, ρ).*

   This is done by induction on the number of disjunctive attacks in (S, ρ). We choose a disjunctive attack and do a case analysis of "imaginary" options, (being option (a), (b,i), (b,ii) and (b,iii) below). With each such option we associate a family $\mathbb{F}$ (option) of networks with a lesser number of disjunctive attacks. Each member of each family will yield some legitimate λ by the induction hypothesis, and the totality of these λ are the legitimate extensions for (S, ρ).

   So let us begin:

Base Case. *There are no disjunctive attacks, but there are constraints λ_1, requiring values from {in, out, und, wave}. Use principles (TC1)–(TC4) of Remark 3.4 to get the extensions, if possible.*

Inductive Case. *There are disjunctive attacks and there are constraints λ_i. In this case we choose one disjunctive attack. Define the case analysis below and define the sets $\mathbb{F}$ (case number). Any λ found by the inductive hypothesis for any element of these sets will do for our (S, ρ).*

So let us begin the inductive case: Let $x\rho\{h_1, \ldots, h_k\}$ as in Figure 2.

We distinguish two cases for the Talmudic complete extension λ.

(a) **case of collapse** *In this case the attack of x on $\{h_1, \ldots, h_k\}$ does collapse to one of the attacks $x \twoheadrightarrow h_i$, for some i.*

Therefore we define the legitimate λ for (S, ρ) as any legitimate λ for $\mathbb{F}$ (case (a)) $= \{(S, \rho_i)|$ where $\rho_i = (\rho - \{(x, \{h_1, \ldots, h_k\})\}) \cup \{(x, \{h_i\})\}\}$ respecting the constraints λ_1.

(b) **case of no collapse** *In this case we distinguish three cases.*

  i. *x is out. In this case there is no attack and we let the legitimate λ for (S, ρ) to be any one of the legitimate λ of $\mathbb{F}$ (case (b,i)) $= \{(S, \rho_i)\}$ of case (a) but with the additional constraint to λ_1 being the constraint $x = $ out$\}$.*

  ii. *x is in or $x = $ wave. In this case there is no collapse and so we have the additional constraints for λ_1 being $h_1 = h_2 = \ldots = h_k = $ wave. So we let the legitimate λ for this case for (S, ρ) to be any legitimate λ for the network $\mathbb{F}$ (case (b,ii)) $= \{(S, \rho')$ where $\rho' = \rho - \{(x, \{h_i, \ldots, h_k\})\}\}$ under the constraint λ_1 augmented by the additional constraints $x = $ in or $x = $ wave ,respectively and $h_i = $ wave for $i = 1, \ldots, k\}$.*

  iii. *x is und. In this case we look at (S, ρ') as in case (ii), with the additional constraints to λ_1 being the constraint $x = h_1 = \ldots = h_k = $ und.*

Example 3.6. *Let us see what the Talmud would do with Figure 10.*

Here we have only one disjunctive attack $x\rho\{a, b\}$ for which we know $x = $ in because x is not attacked. So there are two possibilities for this attack.

1. *The attack collapses and so $x \twoheadrightarrow \{a, b\}$ is to be replaced either by $x \twoheadrightarrow a$ or by $x \twoheadrightarrow b$, giving rise to Figure 11 or Figure 12.*

2. *The attack does not collapse, giving rise to Figure 13 with the constraints shown.*

   So the possible extensions according to Remark 3.4 are:

Disjunctive Attacks in Talmudic Logic

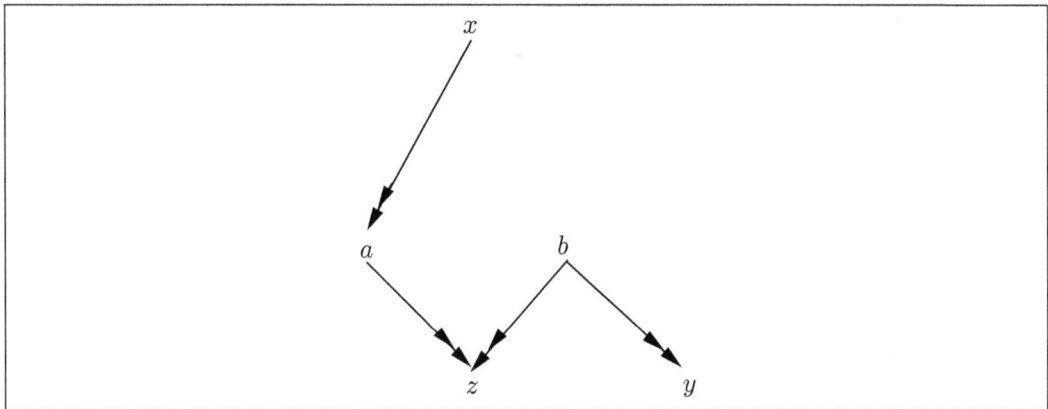

Figure 11

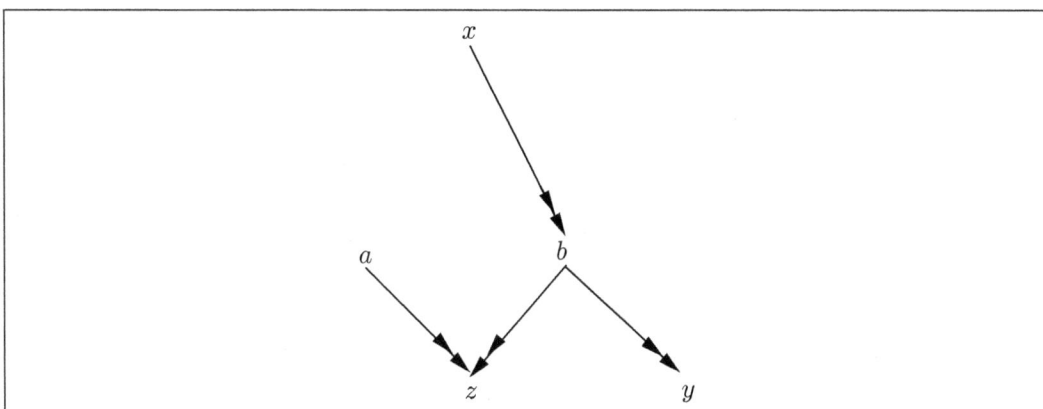

Figure 12

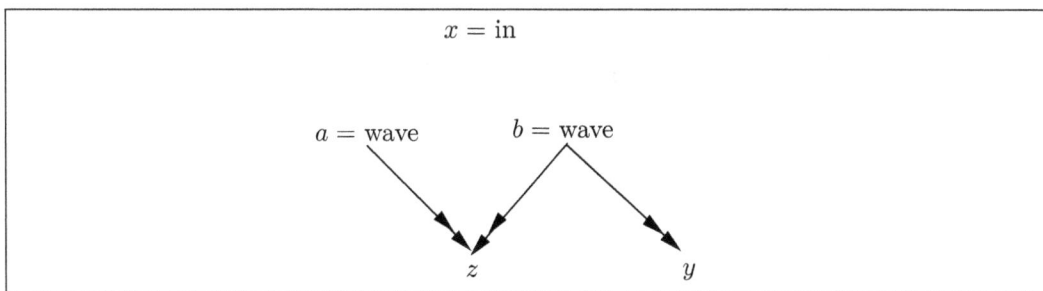

Figure 13

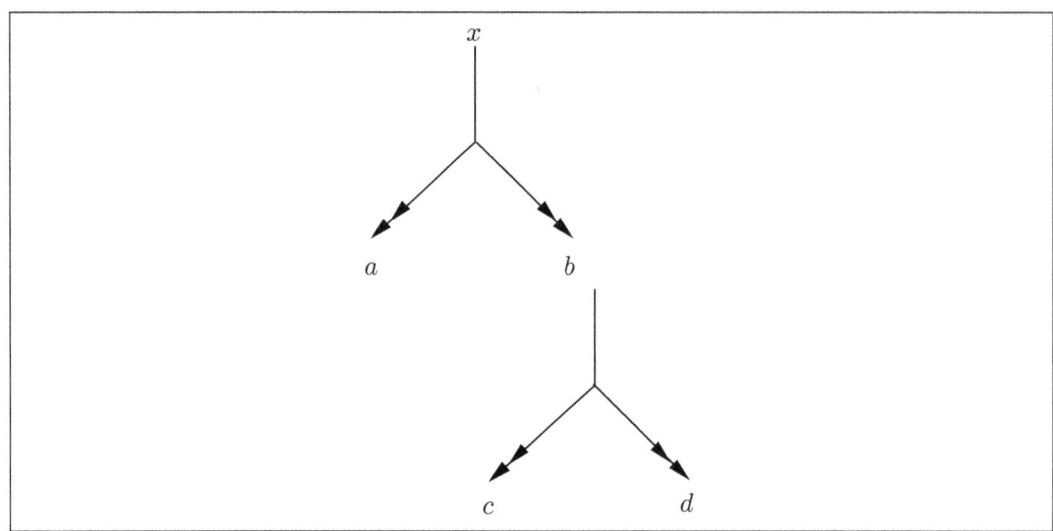

Figure 14

$$\lambda_1 : x = \text{ in}, \ a = \text{ out}, b = \text{ in}, z = y = \text{ out}$$
$$\lambda_2 : x = \text{ in}, \ a = \text{ in}, b = \text{ out}, z = \text{ out}, y = \text{ in}$$
$$\lambda_3 : x = \text{ in}, \ a = b = z = y = \text{ wave}.$$

Example 3.7. *Consider the network of Figure 14.*

Let us agree that $x\rho\{a,b\}$ does collapse while $b\rho\{c,d\}$ does not collapse.

The extensions are the following, calculated intuitively.

$$\lambda_1 : x = \text{in}, \ a = \text{ out}, \ b = \text{ in}, \ c = d = \text{ wave}$$
$$\lambda_2 : x = \text{in}, \ a = \text{ in}, \ b = \text{ out}, \ c = d = \text{ in}$$

Let us now follow our inductive procedure of Remark 3.5 and let us start inductively from $b\rho\{c,d\}$. We get four options, as seen in Figures 15, 16, 17 and 18. The constraints are written in the figures.

For each of the Figures 15–18 we deal with the attack $x\rho\{a,b\}$. These split into two figures each. One with the attack of x on a and one with the attack of x on b.

Some of these will not be possible.

Here are the Figures:

We see that the inductive procedure gave us λ_1 and λ_2 as we expected.

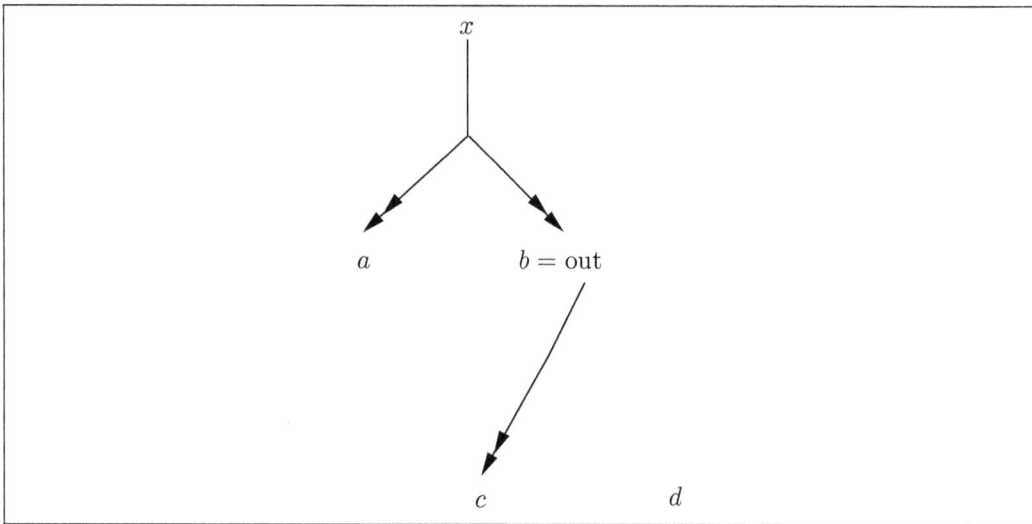

Figure 15

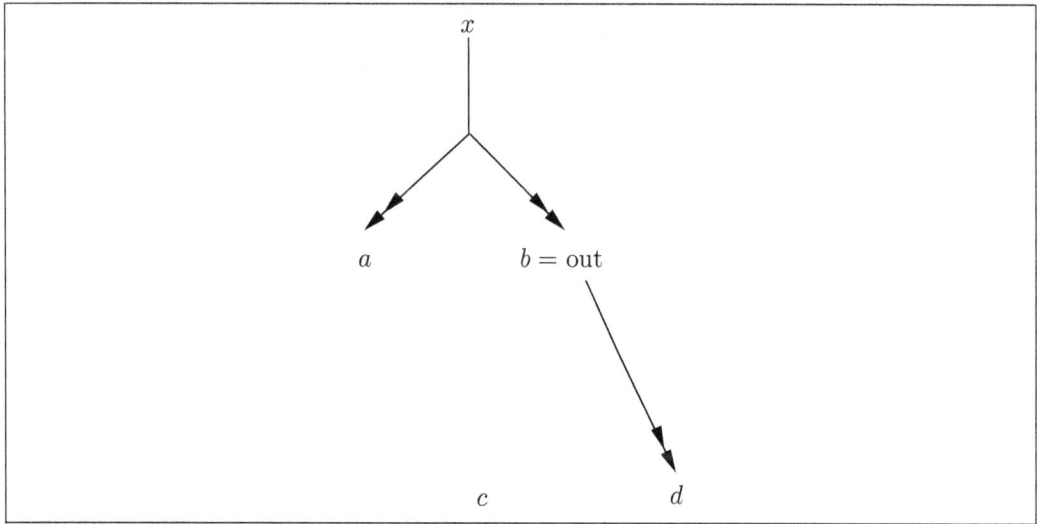

Figure 16

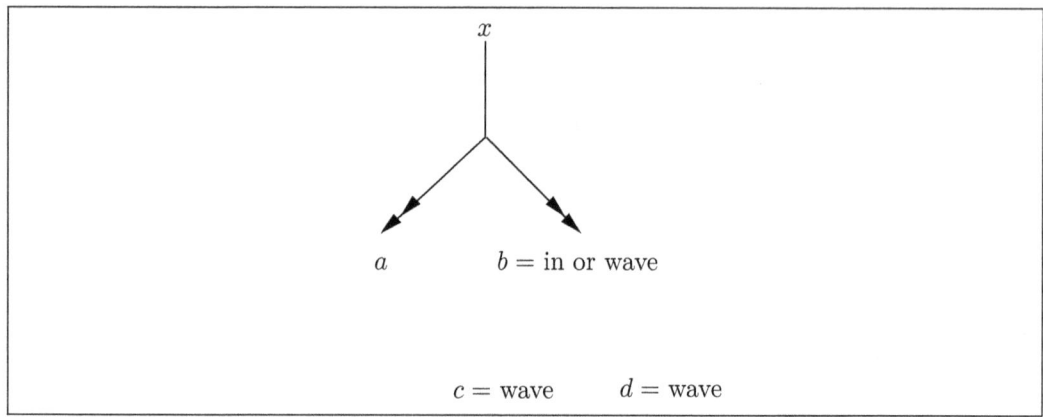

Figure 17

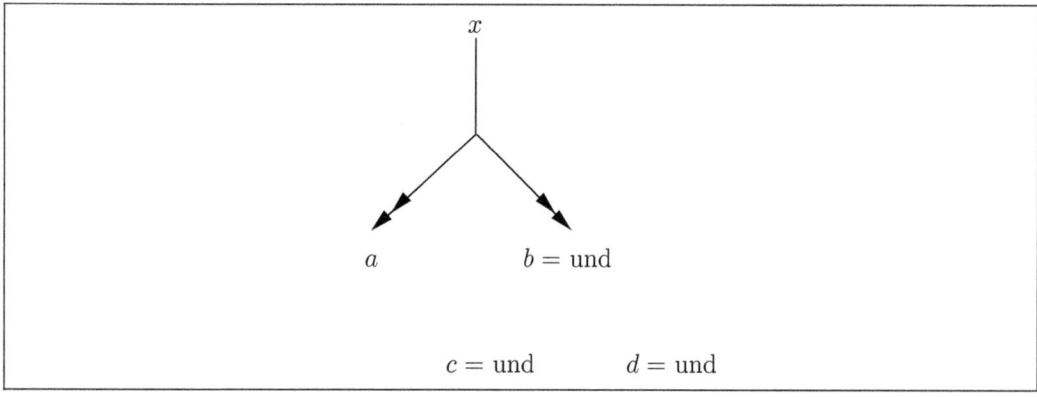

Figure 18

4 Using Bochman's collective argumentation

In his paper [3], Bochman considered conjunctive disjunctive attacks of the form $G \twoheadrightarrow H$, where both G and H are subsets of S. The intended meaning of $G \twoheadrightarrow H$ is that if all embers of G are in, then at least one member of H is out. The treatment of this notion is straightforward (see [6] and Bochman [3]) using axiomatic properties on $G \twoheadrightarrow H$ to characterise various types of semantics. For example, he considered the following axioms:

Montonicity. If $G \twoheadrightarrow H$ then $G \cup G' \twoheadrightarrow H \cup H'$.

Symmetry. If $G \twoheadrightarrow H_1 \cup H_2$ then $G \cup H_1 \twoheadrightarrow H_2$.

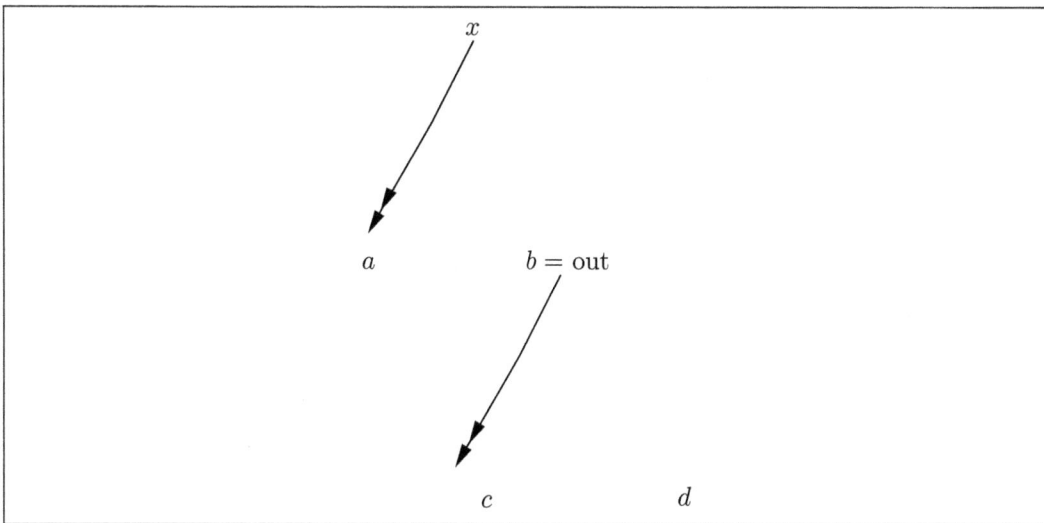

Figure 19: Not possible.

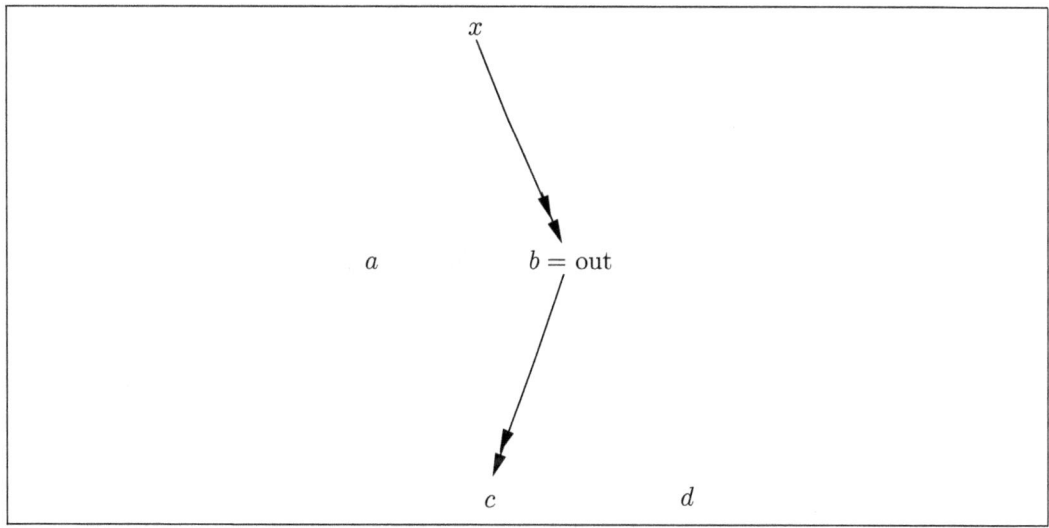

Figure 20: Possible, gives λ_2.

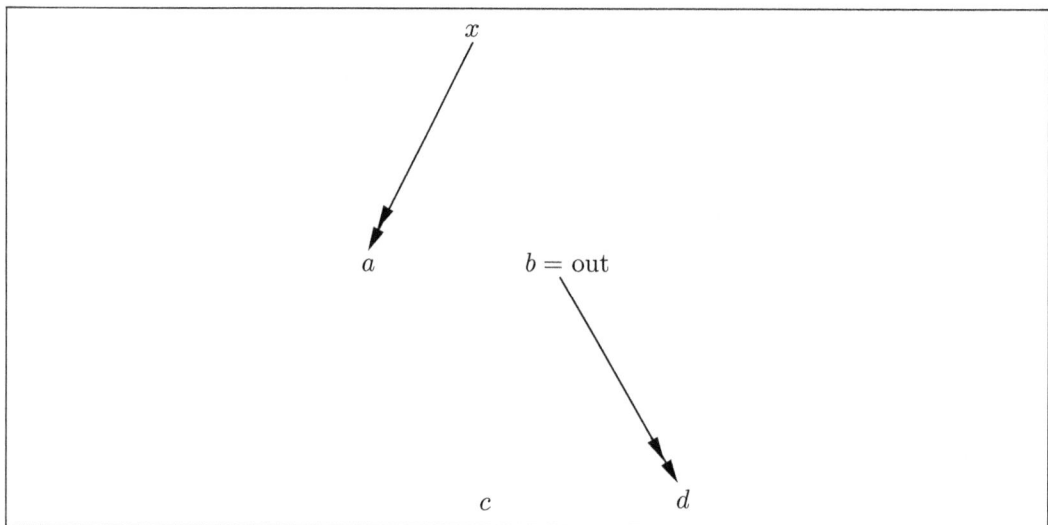

Figure 21: Not possible.

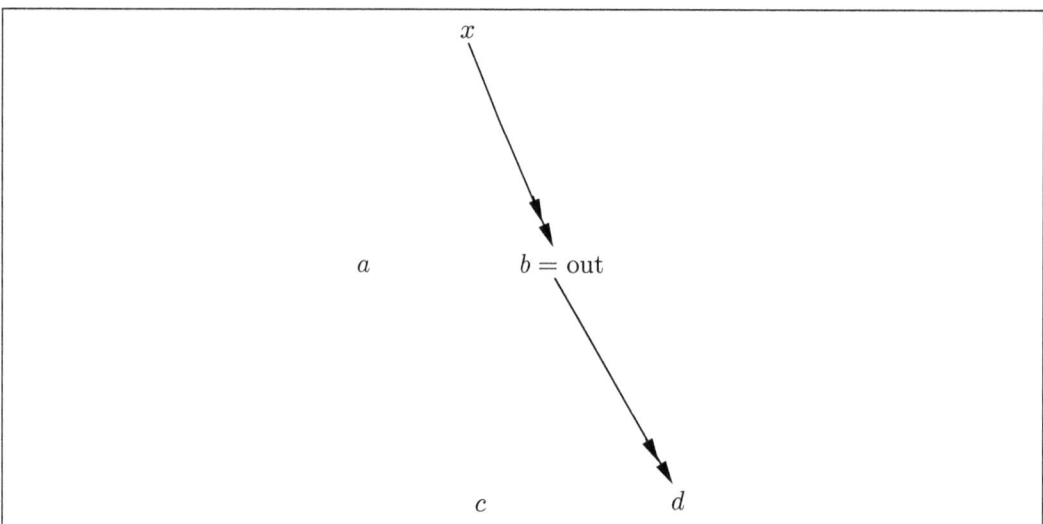

Figure 22: Possible, gives λ_2.

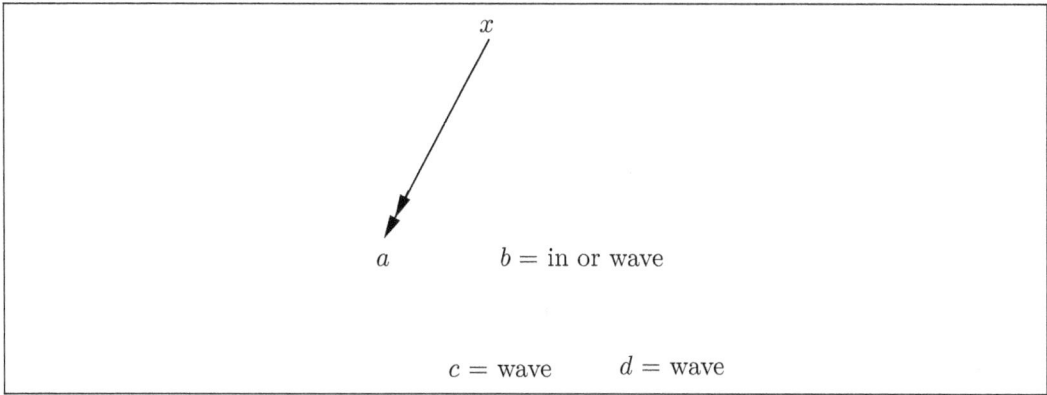
Figure 23: Possible only with $x =$ in. Gives λ_1.

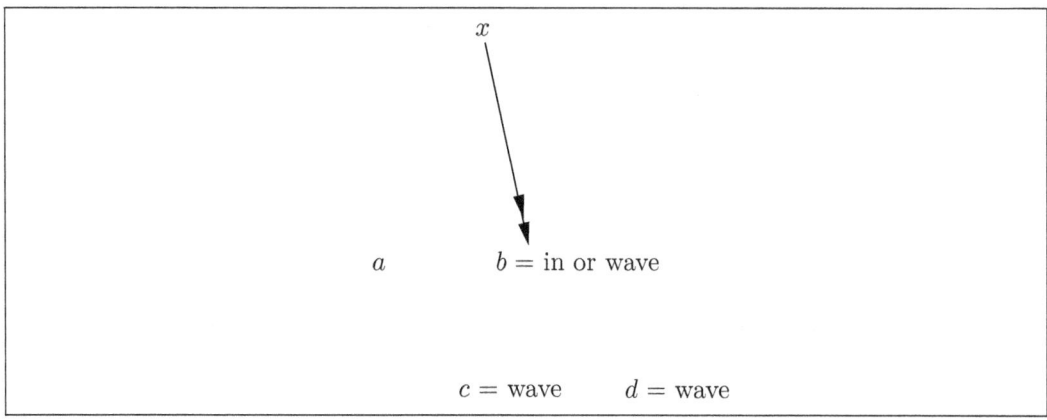
Figure 24: Not possible

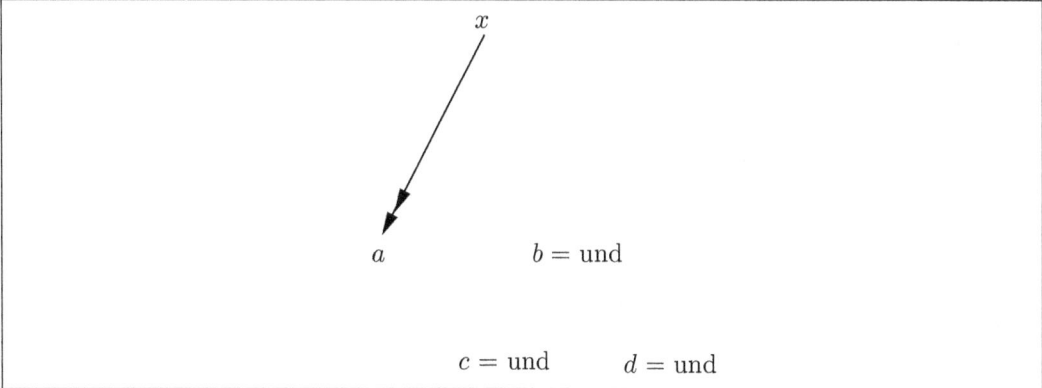
Figure 25: Not possible.

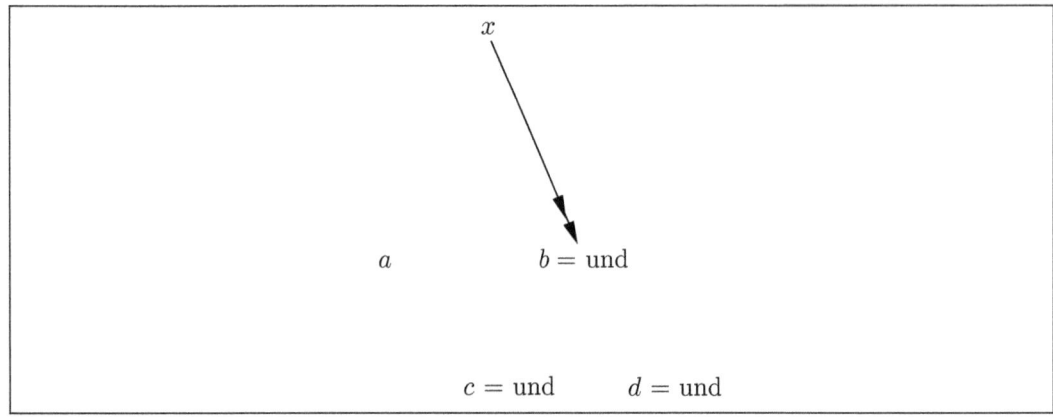

Figure 26: Not possible

Affirmativity. The empty set is not attacked.

Locality. If $G \twoheadrightarrow H_1 \cup H_2$ then $G \twoheadrightarrow H_1$ or $G \twoheadrightarrow H_2$.

Using this axiomatic approach, we want to characterise Talmudic disjunctive attacks. Let us list the properties we need to characterise:

(P1) If $G \twoheadrightarrow H$ then either (a) or (b) holds

  (a) For exactly one $y \in H$ we have $G \twoheadrightarrow \{y\}$. (This is collapse.)

  (b) For none of $H' \subsetneq H$ do we have $G \twoheadrightarrow H'$. This is quantum superposition.)

(P2) If $G \twoheadrightarrow H$ is a quantum superposition attack on H and for some $H' \neq \emptyset$ such that $H' \subseteq H$ we have that $H' \twoheadrightarrow K$ then $H' \twoheadrightarrow K$ is also a quantum superposition attack and furthermore $G \twoheadrightarrow (H - H') \cup K$ also holds as a quantum superpsoition attack.

Figure 27 explains the idea in terms of labels. Any attacker with a wave label propagates this label to its targets.

Note that (P2) can be better understood in positive terms. If $G \to H$ and $H' \to K$ and $H' \subseteq H$, then $G \to (H - H') \cup K$.

The following Bochman-style axioms can characterise (P1) and (P2).

(AP1) If $G \twoheadrightarrow H$ then $H \neq \emptyset$ and

$$\bigvee_{y \in H} \{[G \twoheadrightarrow \{y\}] \wedge \bigwedge_{z \neq y} \neg G \twoheadrightarrow \{z\}\} \vee (\bigwedge_{H' \subsetneq H} (\neg(G \twoheadrightarrow H')))$$

810

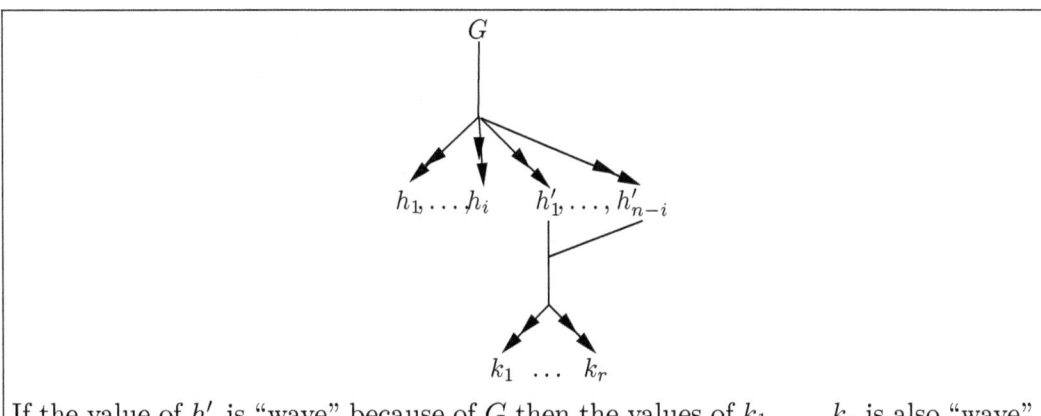

If the value of h'_j is "wave" because of G then the values of $k_1, \ldots, k_r$ is also "wave".

Figure 27

(AP2) If $G \twoheadrightarrow H$ and $\varnothing \neq H' \subsetneq H$ and $H' \twoheadrightarrow K$ then $G \twoheadrightarrow (H - H') \cup K$ and $\bigwedge_{K' \subsetneq K}(\neg(H' \twoheadrightarrow K'))$ and $\bigwedge_{K' \subsetneq (H-H') \cup K}(\neg(G \twoheadrightarrow K'))$.

5 Conclusion and discussion

We saw that Talmudic disjunctive attacks require four values, {in, out, und, wave} and differs from [6] in two senses:

1. The attack on a target set H can turn the target set into having the value "wave" for quantum like superposition. This value is then propagated further by the members of the target set when they attack further targets.

2. When attacking a target set H the attack can collapse to attacking a single $y \in H$. Note and compare that the disjunctive attacks in general can collapse to attacking a subset $\varnothing \neq G \subseteq H$.

   In our case the options for the attack on a set H is either an attack on a single member of H or rendering H into a wave quantum superposition state.

3. We note here, that in view of references [6] and [3], one of our referees commented as follows:

   "Two things should be distinguished here. First, the quantum superposition idea, taken by itself, is more or less comprehensible and internally coherent. The only question is whether it is relevant to argumentation, and it is here that I have my reservations. An argument may collectively attack a set of arguments H just because it disproves one of the joint conclusions of H. This is not a quantum phenomenon, and different arguments in H can still be separated by other arguments, and by different attacks they (separately) create against further arguments. So, if your interpretation insists on existence of a special 'superposition' of a set of arguments, it ought to be represented as an entirely new connective for combining arguments, over and above the existing argumentation machinery."

We comment in return that adding a connective on a set H which turns it into a quantum state is too strong a move. In our system a set H may turn into a quantum state only when attacked. It is the nature of the attack that causes the quantum state. We can certainly investigate a connective, say $\mathbb{Q}(H)$, which turn H into a quantum state but it will be different

(a) A disjunctive attack cannot turn a single point set into a quantum state, but the connective $\mathbb{Q}$ can do that (and make this point propagate the "wave" value).

(b) Adding $\mathbb{Q}$ yields a different logic. (Certainly worthy of investigation.)

4. Another referee brought to our attention the works of Andrew Schumann, [7, 8], connecting Talmudic Logic to parallel computation. The referee spent a lot of effort going through our papers commenting how the quantum view can, and maybe should, be replaced by a parallel computation view. Let us respond to the referee's proposal, i.e. comment on the connection between the phenomena that we describe and parallel computation. Parallel computation describes computational processes that are carried out in parallel. It focuses on processes that cannot be done serially. The focus of the logical problems in our case (a man that marries one of two women) is not connected to the serial question. If the problem was that one cannot decide the state of the one without previously knowing the state of the other or vice versa, the question of serialism would have been relevant. But our problem is totally different. The state cannot be fully decided, even if we do the computation serially. The fact that one of the women is married prevents the marriage of the other, without any connection to the order of computation. It is therefore a problem of Quantum Logic (intertwining of states) and not Parallel Computation Logic. Said another way, in our case there is a complex interaction between the two channels of the problem (like the interaction between distant particles in an ERP experiment). This is the focus of our investigations, and not just the

existence of two parallel channels. The logic of the created state is what we discuss, and this logic is Quantum Logic. We have no interest in how to do the computation that helps us reach the conclusion that this is indeed the state.

References

[1] M. Abraham, I. Belfer, D. Gabbay, and U. Schild. *Principles of Talmudic Logic*. College Publications, 2013.

[2] M. Abraham, I. Belfer, D. Gabbay, and U. Schild. *Fuzzy Logic and Quantum States in Talmudic Reasoning*. College Publications, 2015.

[3] A. Bochman. Collective argumentation and disjunctive logic programming. *J. Logic and Computation*, 13(3):405–428, 2003.

[4] M. Caminada and D. M. Gabbay. A logical account of formal argumentation. *Studia Logica*, 93(2-3):109–145, 2009.

[5] R. Feynman, R. Leighton, and M. Sands. *The Feynman Lectures on Physics*, volume III, chapter 1. Addison Wesley, Reading, 1965.

[6] D. Gabbay and M. Gabbay. Disjunctive attacks in argumentation networks, part 1. *Logic Journal of the IGPL*, 24(2): 186–218, 2016. doi: 10.1093/jigpal/jzv032 First published online: September 10, 2015.

[7] Andrew Schumann. Preface. *History and Philosophy of Logic*, 32(1):1–8, 2011. doi = 10.1080/01445340.2010.506079.

[8] Andrew Schumann. Qal wa-omer and theory of massive-parallel proofs. *History and Philosophy of Logic*, 32(1):71–83, 2011. doi = 10.1080/01445340.2010.506104.

More Modal Semantics Without Possible Worlds

Hitoshi Omori
Department of Philosophy
Kyoto University, Japan.
`hitoshiomori@gmail.com`

Daniel Skurt
RUB Research School & Department of Philosophy II
Ruhr-Universität Bochum, Germany
`daniel.skurt@rub.de`

Abstract

In one of his papers, John Kearns developed a semantics without possible worlds for the normal modal logics **T**, **S4**, and **S5** and even though his work challenged the orthodox view of Kripkean semantics, his work has never been widely recognized. This probably was due to the complexity of his non-deterministic hierarchic semantics. But contemporary developments in the area of non-deterministic semantics make his work most likely understandable for a bigger audience. In this paper we first try to clarify Kearns' original approach within the context of non-deterministic semantics and give proofs of soundness and completeness with respect to Hilbert style calculi. Then, we simplify some of Kearns results and reveal its problematic aspects. Finally, we generalize the result to cover more modal logics such as **K**, **KD** and **KTB**.

Hitoshi Omori was a Postdoctoral Fellow for Research Abroad of the Japan Society for the Promotion of Science (JSPS) at the time of submission, and now he is a Postdoctoral Research Fellow of JSPS. Daniel Skurt was supported by the Ruhr University Research School PLUS, funded by Germany's Excellence Initiative [DFG GSC 98/3]. We would like to thank Thomas Ferguson, Rohan French, David Ripley, Christian Straßer and Heinrich Wansing for helpful discussions and/or comments on an earlier draft. Furthermore, we would like to thank Uwe Meinel and Michael De for proofreading our English. Finally, we would like to thank the anonymous referees for correcting our mistakes, and pointing out some important points which were missing in our draft.

1 Introduction

In [11], John Kearns develops some semantics without possible worlds for modal logics **T**, **S4** and **S5**, based on his negative opinion about possible worlds. For example, Kearns states that "I do not think there are such things as possible worlds, or even that they constitute a useful fiction."([11, p.86])

Our main concern in this paper, however, is purely technical, even though our interest is philosophical.[1] Indeed, if we pay our attention to Kearns' result under such a perspective, then, for example, we find that the basic idea of the framework now known as non-deterministic semantics is actually already used by Kearns. Non-deterministic semantics was first systematically developed by Arnon Avron and Iddo Lev in [1], and has been applied to a wide range of systems of nonclassical logic.[2] Note that Yuri V. Ivlev also considers non-deterministic semantics for modal logics in [9], although he is not dealing with normal modal logics but only fragments *without* the rule of necessitation.[3]

Based on these, the aims of this paper are twofold. First, we clarify the original proof of Kearns, and simplify some of the results. Second, we generalize the result to cover more modal logics such as **K**, **KD** and **KTB**. We will also deal with semantic consequences that Kearns did not investigate, provide proof theories, and prove their soundness and completeness.

The paper is organized as follows. In §2, we present the semantics and proof theory for the modal logic **T**, and in §3 we prove the soundness and completeness regarding the Hilbert-style calculi. These will be followed by §4 in which the soundness and completeness results are extended to the modal logics **S4** and **S5** following the strategy of Kearns. Here, we will also show that the semantics for **S4** and **S5** may be simplified. §5 is devoted to the new results extending the result of Kearns to modal logics **K**, **KD** and **KTB** which are well-known normal modal logics. We will then conclude the paper by summarizing the results, and identifying future directions. We will also briefly remark on the philosophical issues related to the semantics established by Kearns.

Although Kearns deals with quantifiers as well, we will restrict our attention to propositional connectives since they already give rise to plenty of interesting problems.[4]

---

[1] For a follow up paper on the philosophical side of the semantics by Kearns, see [12].

[2] For a survey on non-deterministic semantics, see [2].

[3] Ivlev is not referring to Kearns' result in [9], but he refers to Kearns' paper in [10].

[4] Note that Kearns himself notes, in [12, p.299], that there is an error in his completeness proof for quantified modal logic, i.e. Lemma 3 of [11, p.85], which is corrected in [12]. For a recent work on quantification in non-deterministic semantics in general, see [7].

Finally, at the very end of our study, we discovered that there are some similar attempts in generalizing the result of Kearns by Luis Fariñas del Cerro, Marcelo E. Coniglio and Newton Peron (cf. [6]). Moreover, after our first submission, we noticed a paper [5] published by the same authors. Our results were obtained independently of [6, 5], but since the present paper shares the topic with [5], there will be some overlap. However, in order to keep the paper self-contained as much as possible, we decided to keep the repetition of some of the results already reported in [5]. We only note that the following three results are presented here for the first time and deserve to be highlighted:

- simplification of the semantics of Kearns for **S4** and **S5** (cf. §4.4);
- introduction of Kearns' style semantics for **K** and a proof of its soundness and completeness with respect to the standard Hilbert-style calculi (cf. §5.1);
- observation of an error in the results of Ivlev (cf. §3.3).

2 Semantics and proof theory for T

Our language $\mathcal{L}$ consists of a finite set $\{\neg, \Box, \Diamond, \rightarrow\}$ of propositional connectives and a countable set **Prop** of propositional parameters. Furthermore, we denote by **Form** the set of formulas defined as usual in $\mathcal{L}$. We denote formulas of $\mathcal{L}$ by A, B, C, etc. and sets of formulas of $\mathcal{L}$ by Γ, Δ, Σ, etc.

2.1 Semantics

First, we present the semantics by making use of the general framework of non-deterministic semantics. This will serve as the base of the semantics developed by Kearns.

Definition 1. A **T**-*Nmatrix* for $\mathcal{L}$ is a tuple $M = \langle \mathcal{V}, \mathcal{D}, \mathcal{O} \rangle$, where:

(a) $\mathcal{V} = \{\mathbf{T}, \mathbf{t}, \mathbf{f}, \mathbf{F}\}$,
(b) $\mathcal{D} = \{\mathbf{T}, \mathbf{t}\}$,
(c) For every n-ary connective $*$ of $\mathcal{L}$, $\mathcal{O}$ includes a corresponding n-ary function $\tilde{*}$ from $\mathcal{V}^n$ to $2^{\mathcal{V}} \setminus \{\emptyset\}$ as follows (we omit the brackets for sets):

A	$\neg A$	$\Box A$	$\Diamond A$	$A \rightarrow B$	**T**	**t**	**f**	**F**
T	F	T,t	T,t	**T**	T	t	f	F
t	f	f,F	T,t	**t**	T	T,t	f	f
f	t	f,F	T,t	**f**	T	T,t	T,t	t
F	T	f,F	f,F	**F**	T	T	T	T

A *legal-**T**-valuation* in a **T**-Nmatrix M is a function $v : \mathsf{Form} \to \mathcal{V}$ that satisfies the following condition for every n-ary connective $*$ of $\mathcal{L}$ and $A_1, \ldots, A_n \in \mathsf{Form}$:

$$v(*(A_1, \ldots, A_n)) \in \tilde{*}(v(A_1), \ldots, v(A_n))$$

Remark 2. Note that the four values **T**, **t**, **f** and **F** intuitively represent necessary truth, contingent truth, contingent falsity and necessary falsity (or impossibility). This is reflected in the proof of Lemma 24 which is the key for the completeness.

Definition 3. A is a *legal-**T**-consequence* of Γ ($\Gamma \models_{\text{legal-}\mathbf{T}} A$) iff for all legal-**T**-valuations v, if $v(B) \in \mathcal{D}$ for all $B \in \Gamma$ then $v(A) \in \mathcal{D}$. In particular, A is a *legal-**T**-tautology* iff $v(A) \in \mathcal{D}$ for all legal-**T**-valuations v.

Remark 4. In the original paper of Kearns, the legal-**T**-consequence relation is not defined. However, taking into account the recent development of the non-deterministic semantics, legal-**T**-consequence should be of interest from a technical perspective, and thus we will deal with such relations in the following together with the semantic consequence relation that characterizes the well-known modal logics.

Let us now observe why this legal-**T**-consequence is not sufficient as a semantics for the modal logic **T** through an example.

Example 5. $A \to A$ is a legal-**T**-tautology, but $\Box(A \to A)$ is *not* a legal-**T**-tautology. Indeed, we obtain the following table. Basically, the calculation goes as in many-valued semantics, but now there are some cases that will split as follows.

A	$A \to A$
T	**T**
t	**T**
t	**t**
f	**T**
f	**t**
F	**T**

This shows that $A \to A$ is a legal-**T**-tautology, since $v(A \to A) \in \mathcal{D}$ for all legal-**T**-valuations v. However, $\Box(A \to A)$ is *not* a legal-**T**-tautology since in the third line of the table, when A is assigned the value **t**, then $A \to A$ takes the value **t** as well, and this gives us the result that $\Box(A \to A)$ will take one of the values **f**, **F** which are not designated.

Since we want formulas such as $\Box(A \to A)$ to be validated, we need a different kind of valuations to define the semantic consequence relation. The next definition introduces those valuations which are the remarkable finding of Kearns.

Definition 6. Let v be a function $v : \mathsf{Form} \to \mathcal{V}$. Then,

- v is a *0th-level-**T**-valuation* if v is a legal-**T**-valuation.
- v is a *$m+1$st-level-**T**-valuation* iff v is an mth-level-**T**-valuation and assigns the value **T** to every sentence A if $v'(A) \in \mathcal{D}$ for all mth-level-**T**-valuations v'.

Based on these, we define v to be a **T**-*valuation* iff v is mth-level-**T**-valuation for every $m \geq 0$.

Definition 7. A is a **T**-*tautology* ($\models_{\mathbf{T}} A$) iff $v(A) = \mathbf{T}$ for all **T**-valuations v.

Remark 8. Kearns defined **T**-tautology by collecting formulas that satisfy $v(A) \in \mathcal{D}$ for all **T**-valuations v, but we can replace the condition by a more strict condition as above.

Example 9. Let us illustrate how the 1st-level-**T**-valuations are obtained from the legal-**T**-valuation by using the example above. Basically, we need to rule out some of the legal-**T**-valuations, and this is done by looking at the legal-**T**-tautologies. In the case above, we had six legal-**T**-valuations, but two of them presented in the third and the fifth line will not be counted as 1st-level-**T**-valuations, since they assign the value **t** to the legal-**T**-tautology $A \to A$. And if we focus on 1st-level-**T**-valuations, then it is clear that $\Box(A \to A)$ is a **T**-tautology, as desired. Note that we cannot stop at the 1st level since if we consider the formula $\Box\Box(A \to A)$, then this will *not* be true for some of the 1st-level-**T**-valuations. If we write down the valuations for $\Box\Box(A \to A)$ up to the 3rd level, then it will look as follows. (For the sake of saving space, we have adopted another way of writing down the tables in the following.)

| 0th-level | | | | | 1st-level | | | | | 2nd-level | | | | | 3rd-level | | | | |
$\Box\Box$	$\Box$	$(A$	$\to$	$A)$	$\Box\Box$	$\Box$	$(A$	$\to$	$A)$	$\Box\Box$	$\Box$	$(A$	$\to$	$A)$	$\Box\Box$	$\Box$	$(A$	$\to$	$A)$
T,t,f,F	T,t	T	T	T	T,t,f,F	T,t	T	T	T	T,t	T	T	T	T	T	T	T	T	T
T,t,f,F f,F	T,t f,F	t t	T	t	T,t,f,F	T,t	t	T	t	T,t	T	t	T	t	T	T	t	T	t
T,t,f,F f,F	T,t f,F	f f	T	f	T,t,f,F	T,t	f	T	f	T,t	T	f	T	f	T	T	f	T	f
T,t,f,F	T,t	F	T	F	T,t,f,F	T,t	F	T	F	T,t	T	F	T	F	T	T	F	T	F

At the 0th level, there are 24 relevant valuations.[5] Then, by looking at the formula $A \to A$ and by the definition of the mth-level-**T**-valuation, we rule out the valuations of the third and the fifth line which assign **t** to $A \to A$ to obtain the 1st-level-**T**-valuation. Now, in order to obtain the 2nd-level-**T**-valuation, we look at $\Box(A \to A)$ which takes the designated values for all 1st-level-**T**-valuations, and this time we rule out valuations that assign **t** to $\Box(A \to A)$ to obtain the 2nd-level-**T**-valuation.

---

[5]If the value **f** is assigned to $\Box(A \to A)$, $\Box\Box(A \to A)$ can get the values **F** or **f**. The same counts if the value **F** is assigned to $\Box(A \to A)$.

And finally, we look at $\Box\Box(A\to A)$ which takes the designated values for all 2nd-level-**T**-valuations, and this time we rule out 4 valuations that assign **t** to $\Box\Box(A\to A)$ to obtain the 3rd-level-**T**-valuation. Therefore, taking into account the fact that we can not reduce iterated modalities in the modal logic **T**, we will need the whole hierarchy of valuations. However, one may easily guess that we might not need the whole hierarchy in modal logics in which modalities are reduced to a certain extent. For the details, see §4.4.

Remark 10. Although we introduced $\to$ as the only binary connective, for the purpose of saving labor, it is possible to define conjunction and disjunction in terms of conditional and negation in the usual manner. For example, the truth-table for conjunction becomes as follows:

$A\land B$	**T**	t	f	**F**
T	**T**	t	f	**F**
t	t	t	F,f	**F**
f	f	F,f	F,f	**F**
F	**F**	**F**	**F**	**F**

Now, since $A\land B := \neg(A\to\neg B)$ one would expect that sentences with and without $\land$ behave exactly the same way if they are semantically equivalent. For example, consider the 1st-level-**T**-valuations for $A\land\neg A$ and $\neg(A\to A)$. Then, one might think that we obtain the following tables:

A	$A\land\neg A$
T	**F**
t	**F**
t	f
f	**F**
f	f
F	**F**

A	$\neg(A\to A)$
T	**F**
t	**F**
f	**F**
F	**F**

The tables show the 1st-level-valuations of two equivalent sentences. Note that $A\land\neg A$ has six different valuations, while $\neg(A\to A)$ only has four. This is indeed problematic since if this was the case, then we would have $v(\Diamond(A\land\neg A))\in\mathcal{D}$ for some 1st-level-valuation v, and thus $\Diamond(A\land\neg A)$ would be a satisfiable formula in the modal logic **T** which is obviously not the case since we have $\neg\Diamond(A\land\neg A)$, i.e. $\Box\neg(A\land\neg A)$ is provable in **T**.

As expected, the above table for $A\land\neg A$ is *not* correct. The key is that in order to obtain the 1st-level-**T**-valuation, we need to make sure that *all* legal-**T**-tautologies are assigned the value **T**. In the above case, if we assume that six valuations for $A\land\neg A$ are 1st-level-**T**-valuations and consider the **T**-tautology $\neg(A\land\neg A)$, then we obtain that $\neg(A \wedge \neg A)$ is assigned the value **t** in two cases which are ruled out for the case of $\neg(A \rightarrow A)$, and this needs to be prevented in view of the definition of the mth-level-**T**-valuations. The moral here is in order to obtain the $m+1$st-level-**T**-valuation of a formula we need to take into account the mth-level-**T**-valuations for every sentence A in order to rule out valuations (see Definition 6). This observation imposes some problems when it comes to effective decision procedures. For a further discussion, see Remark 42.

2.2 Proof theory

We now present the proof theory. Although Kearns deployed natural deduction, we will here make use of the Hilbert style system, following [4], which is more standard in the literature of modal logic.

Definition 11. First, the system **T** consists of the following axiom schemata and rules of inference.[6]

$A \rightarrow (B \rightarrow A)$ (Ax1)

$(A \rightarrow (B \rightarrow C)) \rightarrow ((A \rightarrow B) \rightarrow (A \rightarrow C))$ (Ax2)

$(\neg B \rightarrow \neg A) \rightarrow (A \rightarrow B)$ (Ax3)

$\dfrac{A \quad A \rightarrow B}{B}$ (MP)

$\Box(A \rightarrow B) \rightarrow (\Box A \rightarrow \Box B)$ (LK)

$\Box A \rightarrow A$ (LT)

$\Diamond A \rightarrow \neg \Box \neg A$ (LM1)

$\neg \Box \neg A \rightarrow \Diamond A$ (LM2)

$\dfrac{A}{\Box A}$ (RN)

We write $\vdash_\mathbf{T} A$ if there is a sequence of formulas $B_1, \ldots, B_n, A$ ($n \geq 0$), such that every formula in the sequence either (i) is an axiom of **T**; or (ii) is obtained by (MP) or (RN) from formulas preceding it in the sequence. Moreover, we define $\Gamma \vdash_\mathbf{T} A$ iff for a finite subset Γ' of Γ, $\vdash_\mathbf{T} C_1 \rightarrow (C_2 \rightarrow (\cdots (C_n \rightarrow A) \cdots))$ where $C_i \in \Gamma' (1 \leq i \leq n)$.

Second, we define a subsystem of **T**, referred to as $\mathbf{T}^-$, which is obtained by eliminating (RN) and adding the following schemata:

$\Diamond \neg A \rightarrow \neg \Box A$ (LM3)

$\neg \Box A \rightarrow \Diamond \neg A$ (LM4)

$\Box(A \rightarrow B) \rightarrow (\Diamond A \rightarrow \Diamond B)$ (LK1)

$\Diamond(A \rightarrow B) \rightarrow (\Box A \rightarrow \Diamond B)$ (LK2)

$\Box \neg(A \rightarrow B) \rightarrow \Box A$ (LK3)

$\Box \neg(A \rightarrow B) \rightarrow \Box \neg B$ (LK4)

$\Diamond \neg(A \rightarrow B) \rightarrow \Diamond A$ (LK5)

$\Diamond \neg(A \rightarrow B) \rightarrow \Diamond \neg B$ (LK6)

---

[6] Note here that (Ax1)–(Ax3) together with (MP) form a well-known axiomatization of classical propositional logic.

We define $\Gamma \vdash_{\mathbf{T}^-} A$ if there is a sequence of formulas $B_1, \ldots, B_n, A$ ($n \geq 0$), such that every formula in the sequence either (i) is an element of Γ (ii) is an axiom of $\mathbf{T}^-$; or (iii) is obtained by (MP) from formulas preceding it in the sequence.

Remark 12. Note that the system $\mathbf{T}^-$ reflects the relation between proof theory and the 0th-level-$\mathbf{T}$-valuations, i.e. legal-$\mathbf{T}$-valuations. Also note that there are infinitely many subsystems of $\mathbf{T}$ between $\mathbf{T}^-$ and $\mathbf{T}$. These subsystems can be obtained by adding sentences with iterated modalities, e.g. $\Box\Box A$, $\Box\Box\Box A$ or $\Box\Box\Box\Box A$, to $\mathbf{T}^-$.

Proposition 13. The deduction theorem holds for both $\vdash_{\mathbf{T}}$ and $\vdash_{\mathbf{T}^-}$.

Proof. For $\vdash_{\mathbf{T}}$, this is immediate in view of the definition of the consequence relation. For $\vdash_{\mathbf{T}^-}$, the proof runs as in the case for classical logic since we have both (Ax1) and (Ax2) as axioms and (MP) is the only rule of inference. □

3 Soundness and completeness for $\mathbf{T}^-$ and $\mathbf{T}$

We first prove the soundness, and then turn to the completeness result.

3.1 Soundness

The soundness for the legal consequence relation is rather straightforward.

Proposition 14 (Soundness for $\mathbf{T}^-$). If $\Gamma \vdash_{\mathbf{T}^-} A$ then $\Gamma \models_{\text{legal-}\mathbf{T}} A$.

Proof. It suffices to check that all axioms are legal-$\mathbf{T}$-tautologies, and that the rule of inference (MP) preserves the designated value. □

For the soundness of $\mathbf{T}$, we need the following lemma.

Lemma 15. Assume that $\vdash_{\mathbf{T}} A$ and that the length of the proof for A is m. Then, for all mth-level-$\mathbf{T}$-valuations v, $v(A) \in \mathcal{D}$.

Proof. By induction on the length m of the proof for $\vdash_{\mathbf{T}} A$. For the base, case in which $m = 1$, A is one of the axioms. And since axioms are legal-$\mathbf{T}$-tautologies, as shown above, they are also $\mathbf{T}$-tautologies. (Note that by definition, if a sentence is true for all mth-level-$\mathbf{T}$-valuations, then it is also true for all $m + 1$st-level-$\mathbf{T}$-valuations.) For the induction step, assume that the result holds for proofs of the length m, and let $B_1, \ldots, B_m, B_{m+1} (= A)$ be the proof for A. Then, there are the following three cases:

- If A is an axiom, then A is true for all legal-$\mathbf{T}$-valuations, and thus for all $m + 1$st-level-$\mathbf{T}$-valuations as well.

- If A is obtained by applying (MP) to B_i and $B_j (= B_i \to A)$, then by induction hypothesis, B_i and B_j are true for all $\max\{i,j\}$th-level-**T**-valuations, and thus for all mth-level-**T**-valuations. And by the truth table for $\to$, we obtain that A is also true for all mth-level-**T**-valuations. Therefore, A is true for all $m+1$st-level-**T**-valuations as well.
- If A is obtained by applying (RN) to B_i, then by induction hypothesis, B_i is true for all ith-level-**T**-valuations. So, for all $i+1$st-level-**T**-valuations, $\Box B_i$, i.e. A is true. Therefore, A is true for all $m+1$st-level-**T**-valuations as well.

This completes the proof. $\square$

Once we have the lemma, soundness follows immediately.

Proposition 16 (Soundness for **T**). *If $\vdash_\mathbf{T} A$ then $\models_\mathbf{T} A$.*

Proof. Let the length of the proof for A be m. Then, by the above lemma, A is true for all mth-level-**T**-valuations, and thus takes the value **T** for all $m+1$st-level-**T**-valuations. Since **T**-valuations are also $m+1$st-level-**T**-valuations, we obtain that A takes the value **T** for all **T**-valuations, as desired. $\square$

3.2 Completeness

We now move to prove completeness. First, we list some provable formulas that will be used in the following proofs.

Proposition 17. *The following formulas are provable in $\mathbf{T}^-$:*

$$\Box A \to (\neg \Box B \to \neg \Box(A \to B)) \quad (1) \qquad A \to (\neg B \to \neg(A \to B)) \quad (4) \qquad \Box \neg \neg A \to \Box A \quad (7)$$
$$\Box A \to (\Box \neg B \to \Box \neg(A \to B)) \quad (2) \qquad A \to (\neg A \to B) \quad (5) \qquad \Diamond A \to \Diamond \neg \neg A \quad (8)$$
$$(A \to B) \to ((\neg A \to B) \to B) \quad (3) \qquad \Box A \to \Box \neg \neg A \quad (6) \qquad \Diamond \neg \neg A \to \Diamond A \quad (9)$$

Proof. (6), (7), (8) and (9) are provable in view of the axioms (LM1), (LM2), (LM3) and (LM4). Others are rather straightforward in view of footnote 6, and we leave the details to the readers. $\square$

Second, we introduce some notions that will be used in the proofs.

Definition 18. We say that a set Σ of formulas is a *theory* if it is closed under $\vdash$, i.e., if $\Sigma \vdash A$ then $A \in \Sigma$ for all A. A theory Σ is *non-trivial* if for some formula A, $A \notin \Sigma$. Finally, we say that a theory Σ is *consistent* if $A \in \Sigma$ or $\neg A \in \Sigma$, and $A \notin \Sigma$ or $\neg A \notin \Sigma$, for all A.

Definition 19. Let $\Sigma \cup \{B\}$ be a set of formulas. Then Σ is *B-saturated* iff (i) $\Sigma \nvdash B$ and (ii) for all A, if $A \notin \Sigma$ then $\Sigma \cup \{A\} \vdash B$. Finally, Σ is *relatively maximal* iff Σ is B-saturated for some B.

Remark 20. Strictly speaking, we do need to specify the consequence relation in defining B-saturated sets and relatively maximal sets. However, in the following, we will generally omit that since it should be clear from the context. When necessary, we refer to the set as **C**-*relatively maximal* set where **C** is the name of the concerned consequence relation.

We then obtain the following well-known lemmas. As the proofs are standard, we will leave them to the reader.

Lemma 21. If Σ is B-saturated for some B, Σ is a consistent theory, and $B \notin \Sigma$.

Lemma 22. If $\Sigma \nvdash A$, there is a $\Pi \supseteq \Sigma$ such that Π is A-saturated.

Definition 23. Let Σ be a relatively maximal set. Then, we define a function v_Σ from Form to $\mathcal{V}$ as follows:

$$v_\Sigma(B) := \begin{cases} \mathbf{T} & \text{if } \Sigma \vdash \Box B \\ \mathbf{t} & \text{if } \Sigma \nvdash \Box B \text{ and } \Sigma \vdash B \\ \mathbf{f} & \text{if } \Sigma \nvdash \Box \neg B \text{ and } \Sigma \vdash \neg B \\ \mathbf{F} & \text{if } \Sigma \vdash \Box \neg B \end{cases}$$

We need one more lemma which is the key for the completeness result.

Lemma 24. If Σ is a **T**-relatively maximal set then v_Σ is a legal-**T**-valuation.

Proof. Note first that v_Σ is well-defined in view of (LT) and (5). Then the desired result is proved by induction on the number n of connectives.
(Base): for atomic formulas, just note that Σ is a consistent theory which implies that $\Sigma \vdash A$ or $\Sigma \vdash \neg A$.
(Induction step): We split the cases based on the connectives.
Case 1. If $B = \neg C$, then we have the following four cases.

cases	$v_\Sigma(C)$	condition for C	$v_\Sigma(B)$	condition for B i.e. $\neg C$
(i)	**T**	$\Sigma \vdash \Box C$	**F**	$\Sigma \vdash \Box \neg \neg C$
(ii)	**t**	$\Sigma \nvdash \Box C$ and $\Sigma \vdash C$	**f**	$\Sigma \nvdash \Box \neg \neg C$ and $\Sigma \vdash \neg \neg C$
(iii)	**f**	$\Sigma \nvdash \Box \neg C$ and $\Sigma \vdash \neg C$	**t**	$\Sigma \nvdash \Box \neg C$ and $\Sigma \vdash \neg C$
(iv)	**F**	$\Sigma \vdash \Box \neg C$	**T**	$\Sigma \vdash \Box \neg C$

By induction hypothesis, we have the conditions for C, and it is easy to see that the conditions for B i.e. $\neg C$ are provable. Indeed, (iii) and (iv) are obvious, and the others are provable (cf. (6), (7), (8) and (9)).
Case 2. If $B = \Box C$, then we have the following four cases.

Modal Semantics Without Possible Worlds

cases	$v_\Sigma(C)$	condition for C	$v_\Sigma(B)$	condition for B i.e. $\Box C$
(i)	T	$\Sigma \vdash \Box C$	T, t	$\Sigma \vdash \Box C$
(ii)	t	$\Sigma \nvdash \Box C$ and $\Sigma \vdash C$	f, F	$\Sigma \vdash \neg \Box C$
(iii)	f	$\Sigma \nvdash \Box \neg C$ and $\Sigma \vdash \neg C$	f, F	$\Sigma \vdash \neg \Box C$
(iv)	F	$\Sigma \vdash \Box \neg C$	f, F	$\Sigma \vdash \neg \Box C$

By induction hypothesis, we have the conditions for C, and we can see that the conditions for B i.e. $\Box C$ are provable. Indeed, (i) and (ii) are obvious, and the others can be proved by (LT).

Case 3. If $B = \Diamond C$, then we have the following four cases.

cases	$v_\Sigma(C)$	condition for C	$v_\Sigma(B)$	condition for B i.e. $\Diamond C$
(i)	T	$\Sigma \vdash \Box C$	T, t	$\Sigma \vdash \Diamond C$
(ii)	t	$\Sigma \nvdash \Box C$ and $\Sigma \vdash C$	T, t	$\Sigma \vdash \Diamond C$
(iii)	f	$\Sigma \nvdash \Box \neg C$ and $\Sigma \vdash \neg C$	T, t	$\Sigma \vdash \Diamond C$
(iv)	F	$\Sigma \vdash \Box \neg C$	f, F	$\Sigma \vdash \neg \Diamond C$

By induction hypothesis, we have the conditions for C, and we can see that the conditions for B i.e. $\Diamond C$ are provable. Indeed, (i) and (ii) are provable by (LT) and (LM2), (iii) is provable by (LM3), and (iv) is provable by (LM1).

Case 4. If $B = C \to D$, then we have the following 11 cases.

cases	$v_\Sigma(C)$	condition for C	$v_\Sigma(D)$	condition for D	$v_\Sigma(B)$	condition for B i.e. $C \to D$
(i)	F	$\Sigma \vdash \Box \neg C$	any	—	T	$\Sigma \vdash \Box(C \to D)$
(ii)	any	—	T	$\Sigma \vdash \Box D$	T	$\Sigma \vdash \Box(C \to D)$
(iii)	T	$\Sigma \vdash \Box C$	t	$\Sigma \nvdash \Box D$ & $\Sigma \vdash D$	t	$\Sigma \nvdash \Box(C \to D)$ & $\Sigma \vdash (C \to D)$
(iv)	T	$\Sigma \vdash \Box C$	f	$\Sigma \nvdash \Box \neg D$ & $\Sigma \vdash \neg D$	f	$\Sigma \nvdash \Box \neg (C \to D)$ & $\Sigma \vdash \neg (C \to D)$
(v)	T	$\Sigma \vdash \Box C$	F	$\Sigma \vdash \Box \neg D$	F	$\Sigma \vdash \Box \neg (C \to D)$
(vi)	t	$\Sigma \nvdash \Box C$ & $\Sigma \vdash C$	t	$\Sigma \nvdash \Box D$ & $\Sigma \vdash D$	T, t	$\Sigma \vdash C \to D$
(vii)	t	$\Sigma \nvdash \Box C$ & $\Sigma \vdash C$	f	$\Sigma \nvdash \Box \neg D$ & $\Sigma \vdash \neg D$	f	$\Sigma \nvdash \Box \neg (C \to D)$ & $\Sigma \vdash \neg (C \to D)$
(viii)	t	$\Sigma \nvdash \Box C$ & $\Sigma \vdash C$	F	$\Sigma \vdash \Box \neg D$	f	$\Sigma \nvdash \Box \neg (C \to D)$ & $\Sigma \vdash \neg (C \to D)$
(ix)	f	$\Sigma \nvdash \Box \neg C$ & $\Sigma \vdash \neg C$	t	$\Sigma \nvdash \Box D$ & $\Sigma \vdash D$	T, t	$\Sigma \vdash C \to D$
(x)	f	$\Sigma \nvdash \Box \neg C$ & $\Sigma \vdash \neg C$	f	$\Sigma \nvdash \Box \neg D$ & $\Sigma \vdash \neg D$	T, t	$\Sigma \vdash C \to D$
(xi)	f	$\Sigma \nvdash \Box \neg C$ & $\Sigma \vdash \neg C$	F	$\Sigma \vdash \Box \neg D$	t	$\Sigma \nvdash \Box(C \to D)$ & $\Sigma \vdash (C \to D)$

By induction hypothesis, we have the conditions for C and D, and we can see that the conditions for B i.e. $C \to D$ are provable as follows:

- For (i) and (ii), use (LK5) and (LK6) respectively.
- For (iii), $\Sigma \vdash C \to D$ follows immediately by $\Sigma \vdash D$ and (Ax1). For the other half, use (LK).
- For (iv), $\Sigma \vdash \neg(C \to D)$ follows in view of (4). For the other half, by (LK4).
- For (v), just use (2).
- For (vi) and (ix), just use (Ax1).
- For (vii) and (viii), $\Sigma \vdash \neg(C \to D)$ follows in view of (4). For the other half, by (LK3).
- For (x), just use (5).
- For (xi), $\Sigma \vdash C \to D$ follows in view of (Ax1). For the other half, by (LK1).

This completes the proof. $\square$

Remark 25. Note that the above Definition 23 shows us the intuitive reading of the four truth values. That is, $\mathbf{T}, \mathbf{t}, \mathbf{f}$ and $\mathbf{F}$ can be read as necessarily true, contingently true, contingently false and necessarily false.

Based on these, we are now ready to prove the completeness of $\mathbf{T}^-$.

Theorem 1 (Completeness for $\mathbf{T}^-$). *If $\Gamma \models_{\text{legal-T}} A$ then $\Gamma \vdash_{\mathbf{T}^-} A$.*

Proof. We prove the contrapositive. Suppose that $\Gamma \not\vdash_{\mathbf{T}^-} A$. Then by Lemma 22, we have an A-saturated set Σ_0 such that $\Gamma \subseteq \Sigma_0$. In view of Lemma 24, we can define a legal-$\mathbf{T}$-valuation v_{Σ_0} such that $v_{\Sigma_0}(B) \in \mathcal{D}$ for all $B \in \Gamma$ and $v_{\Sigma_0}(A) \notin \mathcal{D}$. Thus we have $\Gamma \not\models_{\text{legal-T}} A$, as desired. □

Remark 26. Ivlev also formulates a very similar system to $\mathbf{T}^-$ in [9, p.116] and [10, p.108] where the system is called $\mathbf{S}_a+$. The main difference lies in his additional rule that "we can replace any number of occurrences of $\neg\neg A$ by A and vice versa." ([9, p.115], in our notation). Unfortunately, this rule leads to a system which is not sound with respect to Ivlev's semantics. See §3.3 for a further discussion.

For the completeness of $\mathbf{T}$, we need one more lemma.

Lemma 27. *Let Γ be a $\mathbf{T}$-relatively maximal set. If v_Γ is a legal-$\mathbf{T}$-valuation, then v_Γ is also an mth-level-$\mathbf{T}$-valuation for every $m \geq 1$, and thus a $\mathbf{T}$-valuation.*

Proof. By induction on m. For the base case, we prove that v_Γ is 1st-level-$\mathbf{T}$-valuation. Let A be a sentence that is true for all legal-$\mathbf{T}$-valuations. Assume, for reductio, that $\not\vdash A$. Then by Lemma 22, there is an A-saturated set Σ. Now let v_Σ be the legal-$\mathbf{T}$-valuation generated by Σ. By the definition of v_Σ, we have that $\Sigma \not\vdash A$, i.e. $v(A) \notin \mathcal{D}$. But this contradicts our assumption that A is true for all legal-$\mathbf{T}$-valuations. Therefore, we have proved that $\vdash A$. Then by (RN), we obtain $\vdash \Box A$. By the definition of v_Γ, we obtain that $v_\Gamma(A) = \mathbf{T}$, as desired.

For the induction step, assume that v_Γ is an mth-level-$\mathbf{T}$-valuation, and let A be a sentence that is true for all mth-level-$\mathbf{T}$-valuations. Assume, for contradiction, that $\not\vdash A$. Then by Lemma 22, there is an A-saturated set Σ. Now let v_Σ be the legal-$\mathbf{T}$-valuation generated by Σ. By induction hypothesis, we have that v_Σ is an mth-level-$\mathbf{T}$-valuation. Moreover, by the definition of v_Σ, we have that $\Sigma \not\vdash A$, i.e. $v(A) \notin \mathcal{D}$. But this contradicts our assumption that A is true for all mth-level-$\mathbf{T}$-valuations. Therefore, we have proved that $\vdash A$. Then by (RN), we obtain $\vdash \Box A$. By the definition of v_Γ, we obtain that $v_\Gamma(A) = \mathbf{T}$, as desired. □

Remark 28. Note that this lemma is not relying on anything specific in the modal logic $\mathbf{T}$, but only needs the rule of necessitation from a proof-theoretical perspective. Therefore, this lemma can be proved for any normal modal logic if the legal valuations are defined appropriately.

And now we are ready to prove completeness for **T**.

Theorem 2 (Completeness for **T**). If $\models_\mathbf{T} A$ then $\vdash_\mathbf{T} A$.

Proof. We prove the contrapositive. Suppose that $\nvdash_\mathbf{T} A$. Then by Lemma 22, we have an A-saturated set Σ_0 such that $\Sigma_0 \nvdash A$. In view of Lemma 24, we can define a legal valuation v_{Σ_0}, and by Lemma 27, this v_{Σ_0} is also a **T**-valuation. And since we have $v_{\Sigma_0}(A) \notin \mathcal{D}$, it is also the case that $v_{\Sigma_0}(A) \neq \mathbf{T}$ (since $v_{\Sigma_0}(A) = \mathbf{T}$ implies that $v_{\Sigma_0}(A) \in \mathcal{D}$) and thus we obtain $\nvDash_\mathbf{T} A$, as desired. □

This completes the proof for the soundness and completeness result for **T**. The completeness for **S4** and **S5** only requires some small changes, as we shall see in the next section. Before turning to that, we make a brief remark on Ivlev's system.

3.3 A discussion on the result of Ivlev

In [9], Ivlev introduces a Hilbert style system called $\mathbf{S}_a+$ which is similar to $\mathbf{T}^-$ and formulated in the same language as in this paper. The main difference lies in an additional rule for substituting double negations. Also, the set of axioms is slightly different, as follows.

Definition 29. The system $\mathbf{S}_a+$ consists of the following axiom schemata and rules of inference.

$A \to (B \to A)$ \hfill (Ax1) \hspace{2em} $\neg \Diamond A \to \Box(A \to B)$ \hfill (AM$_5$)

$(A \to (B \to C)) \to ((A \to B) \to (A \to C))$ \hfill (Ax2) \hspace{2em} $\Box B \to \Box(A \to B)$ \hfill (AM$_6$)

$(\neg B \to \neg A) \to (A \to B)$ \hfill (Ax3) \hspace{2em} $\Box(A \to B) \to (\Box A \to \Box B)$ \hfill (AM$_7$)

$\Box A \to A$ \hfill (AM$_1$) \hspace{2em} $\Box(A \to B) \to (\Diamond A \to \Diamond B)$ \hfill (AM$_{8''}$)

$A \to \Diamond A$ \hfill (AM$_2$) \hspace{2em} $\Diamond(A \to B) \to (\Box A \to \Diamond B)$ \hfill (AM$_9$)

$\neg \Box \neg A \to \Diamond A$ \hfill (AM$_3$) \hspace{2em} $\Diamond B \to \Diamond(A \to B)$ \hfill (AM$_{10}$)

$\Diamond A \to \neg \Box \neg A$ \hfill (AM$_4$) \hspace{2em} $\Diamond \neg A \to \Diamond(A \to B)$ \hfill (AM$_{11}$)

$$\dfrac{A \quad A \to B}{B} \text{ (MP)} \qquad \dfrac{B[A]}{B[\neg\neg A]} \quad \dfrac{B[\neg\neg A]}{B[A]} \text{ (RI)}$$

We write $\vdash_{\mathbf{S}_a+} A$ if there is a sequence of formulas $B_1, \ldots, B_n, A$ ($n \geq 0$), such that every formula in the sequence either (i) is an axiom of $\mathbf{S}_a+$; or (ii) is obtained by (MP) or (RI) from formulas preceding it in the sequence.

Note, that Ivlev formulated the rule (RI) as follows: "we can replace any number of occurrences of $\neg\neg A$ by A and vice versa" ([9, p.115]). But, this means (RI) is a particular substitution rule.

Now, the semantic consequence relation considered by Ivlev is the legal-**T**-consequence introduced in this paper (cf. Definition 3).

Observation 30. $\vdash_{\mathbf{S}_a+} \Box\Box A \to \Box\Box\neg\neg A$ but $\not\models_{\text{legal-}\mathbf{T}} \Box\Box A \to \Box\Box\neg\neg A$.

Proof. For the former, by the tautology '$\Box\Box A \to \Box\Box A$' and (RI). For the latter, take a function $v_0 : \mathsf{Form} \to \mathcal{V}$ such that $v_0(A) = \mathbf{T}$, $v_0(\Box A) = \mathbf{T}$, $v_0(\Box\Box A) = \mathbf{T}$, $v_0(\Box\neg\neg A) = \mathbf{t}$ and $v_0(\Box\Box\neg\neg A) = \mathbf{F}$. Then this is a legal-**T**-valuation by which we obtain $v_0(\Box\Box A \to \Box\Box\neg\neg A) = \mathbf{F} \notin \mathcal{D}$, as desired. □

In view of the above observation, we obtain that the system $\mathbf{S}_a+$ is *not* sound with respect to the semantics.

4 Extending the results to S4 and S5

As observed by Kearns, we obtain the soundness and completeness results for **S4** and **S5** by some simple changes. First we revisit them, and then show that some simplification is possible as well.

4.1 Modifications in semantics and proof theory

Definition 31. An **S4**- and **S5**-*Nmatrix* is obtained by replacing the conditions for $\Box$ and $\Diamond$ as in the following tables 1 and 2 respectively:

A	$\Box A$	$\Diamond A$		A	$\Box A$	$\Diamond A$
T	T	T		T	T	T
t	f, F	T, t		t	F	T
f	f, F	T, t		f	F	T
F	F	F		F	F	F
Table 1 (for **S4**)				Table 2 (for **S5**)		

Based on these, legal valuations, legal semantic consequence relations, valuations and consequence relations are all defined as in the cases for $\mathbf{T}^-$ and $\mathbf{T}$.

Definition 32. The systems **S4** and **S5** are obtained by adding (L4) and (M5) to **T** and **S4** respectively.

$$\Box A \to \Box\Box A \quad (L4) \qquad \Diamond A \to \Box\Diamond A \quad (L5)$$

Moreover, we obtain $\mathbf{S4}^-$ by adding the following to $\mathbf{T}^-$:

$$\Box A \to \Box\Box A \quad (L4) \qquad \Box A \to \Box\Diamond A \quad (L4')$$
$$\Diamond\Diamond A \to \Diamond A \quad (M4) \qquad \Diamond\Box A \to \Diamond A \quad (M4')$$

Finally, we obtain **S5⁻** by adding the following to **S4⁻**:

$$\Diamond A \to \Box \Diamond A \quad (L5) \qquad\qquad \Diamond \Box A \to \Box A \quad (M5)$$

The consequence relations are defined as in Definition 11.

4.2 Soundness

What we need to do is to check the validity of the additional axioms with respect to the semantics which is rather straightforward.

Proposition 33 (Soundness for **S4⁻**, **S5⁻**, **S4** and **S5**). We have the following.
- If $\Gamma \vdash_{\mathbf{S4^-}} A$ then $\Gamma \models_{\text{legal-S4}} A$.
- If $\Gamma \vdash_{\mathbf{S5^-}} A$ then $\Gamma \models_{\text{legal-S5}} A$.
- If $\vdash_{\mathbf{S4}} A$ then $\models_{\mathbf{S4}} A$.
- If $\vdash_{\mathbf{S5}} A$ then $\models_{\mathbf{S5}} A$.

Proof. For **S4⁻**, **S5⁻**, we only need to check that the axioms we added to obtain **S4⁻** and **S5⁻** are legal-**S4**-tautologies and legal-**S5**-tautologies respectively. For **S4** and **S5**, just apply the proof for Proposition 16. □

4.3 Completeness

We now turn to completeness. Again, the changes we need to make are all something expected.

Lemma 34. *If Σ is a **S4**-relatively maximal set then v_Σ is a legal-**S4**-valuation.*

Proof. We only need to take care of $\Box$ and $\Diamond$.
Case 1. If $B = \Box C$, then we have the following four cases.

cases	$v_\Sigma(C)$	condition for C	$v_\Sigma(B)$	condition for B i.e. $\Box C$
(i)	T	$\Sigma \vdash \Box C$	T	$\Sigma \vdash \Box\Box C$
(ii)	t	$\Sigma \nvdash \Box C$ and $\Sigma \vdash C$	f, F	$\Sigma \vdash \neg\Box C$
(iii)	f	$\Sigma \nvdash \Box\neg C$ and $\Sigma \vdash \neg C$	f, F	$\Sigma \vdash \neg\Box C$
(iv)	F	$\Sigma \vdash \Box\neg C$	F	$\Sigma \vdash \Box\neg\Box C$

By induction hypothesis, we have the conditions for C, and we can prove the conditions for B i.e. $\Box C$ as follows. (ii) and (iii) are already covered in **T⁻**. For (i) and (iv), just use (L4) and (M4') respectively.
Case 2. If $B = \Diamond C$, then we have the following four cases.

cases	$v_\Sigma(C)$	condition for C	$v_\Sigma(B)$	condition for B i.e. $\Diamond C$
(i)	T	$\Sigma \vdash \Box C$	T	$\Sigma \vdash \Box \Diamond C$
(ii)	t	$\Sigma \nvdash \Box C$ and $\Sigma \vdash C$	T, t	$\Sigma \vdash \Diamond C$
(iii)	f	$\Sigma \nvdash \Box\neg C$ and $\Sigma \vdash \neg C$	T, t	$\Sigma \vdash \Diamond C$
(iv)	F	$\Sigma \vdash \Box\neg C$	F	$\Sigma \vdash \Box\neg\Diamond C$

By induction hypothesis, we have the conditions for C, and we can prove the conditions for B i.e. $\Diamond C$ as follows. (ii) and (iii) are already covered in $\mathbf{T}^-$. For (i) and (iv), just use (L4') and (M4) respectively. This completes the proof. $\square$

Lemma 35. *If Σ is a $\mathbf{S5}$-relatively maximal set then v_Σ is a legal-$\mathbf{S5}$-valuation.*

Proof. We only need to take care of $\Box$ and $\Diamond$.
Case 1. If $B = \Box C$, then we have the following four cases.

cases	$v_\Sigma(C)$	condition for C	$v_\Sigma(B)$	condition for B i.e. $\Box C$
(i)	T	$\Sigma \vdash \Box C$	T	$\Sigma \vdash \Box\Box C$
(ii)	t	$\Sigma \not\vdash \Box C$ and $\Sigma \vdash C$	F	$\Sigma \vdash \Box\neg\Box C$
(iii)	f	$\Sigma \not\vdash \Box\neg C$ and $\Sigma \vdash \neg C$	F	$\Sigma \vdash \Box\neg\Box C$
(iv)	F	$\Sigma \vdash \Box\neg C$	F	$\Sigma \vdash \Box\neg\Box C$

By induction hypothesis, we have the conditions for C, and we can prove the conditions for B i.e. $\Diamond C$ as follows. Since (i) and (iv) are already covered by $\mathbf{S4}^-$, we only consider (ii) and (iii). For (ii), it is sufficient to prove that $\Sigma \not\vdash \Box\neg\Box C$ implies $\Sigma \vdash \Box C$. Now, $\Sigma \not\vdash \Box\neg\Box C$ implies $\Sigma \vdash \neg\Box\neg\Box C$, i.e. $\Sigma \vdash \Diamond\Box C$ and by (M5) this implies $\Sigma \vdash \Box C$ as desired. For (iii), it is sufficient to prove that $\Sigma \not\vdash \Box\neg\Box C$ implies $\Sigma \not\vdash \neg C$. So, assume $\Sigma \not\vdash \Box\neg\Box C$. Then we have $\Sigma \vdash \neg\Box\neg\Box C$, and so $\Sigma \vdash \Diamond\Box C$. This then implies $\Sigma \vdash C$, i.e. $\Sigma \not\vdash \neg C$, by (M5) and (LT).

Case 2. If $B = \Diamond C$, then we have the following four cases.

cases	$v_\Sigma(C)$	condition for C	$v_\Sigma(B)$	condition for B i.e. $\Diamond C$
(i)	T	$\Sigma \vdash \Box C$	T	$\Sigma \vdash \Box\Diamond C$
(ii)	t	$\Sigma \not\vdash \Box C$ and $\Sigma \vdash C$	T	$\Sigma \vdash \Box\Diamond C$
(iii)	f	$\Sigma \not\vdash \Box\neg C$ and $\Sigma \vdash \neg C$	T	$\Sigma \vdash \Box\Diamond C$
(iv)	F	$\Sigma \vdash \Box\neg C$	F	$\Sigma \vdash \Box\neg\Diamond C$

By induction hypothesis, we have the conditions for C, and we can prove the conditions for B i.e. $\Diamond C$ as follows. Since (i) and (iv) are already covered by $\mathbf{S4}^-$, we only consider (ii) and (iii). For (ii), it is sufficient to prove that $\Sigma \vdash C$ implies $\Sigma \vdash \Box\Diamond C$ but this is easy in the light of (LT) and (L5). For (iii), it is sufficient to prove that $\Sigma \not\vdash \Box\neg C$ implies $\Sigma \vdash \Box\Diamond C$. But this is easy again in the light of (LM2) and (L5). This completes the proof. $\square$

Theorem 3 (Completeness for $\mathbf{S4}^-$, $\mathbf{S5}^-$, $\mathbf{S4}$ and $\mathbf{S5}$). *We have the following.*

- *If $\Gamma \models_{\text{legal-}\mathbf{S4}} A$ then $\Gamma \vdash_{\mathbf{S4}^-} A$.*
- *If $\Gamma \models_{\text{legal-}\mathbf{S5}} A$ then $\Gamma \vdash_{\mathbf{S5}^-} A$.*
- *If $\models_{\mathbf{S4}} A$ then $\vdash_{\mathbf{S4}} A$.*
- *If $\models_{\mathbf{S5}} A$ then $\vdash_{\mathbf{S5}} A$.*

Proof. Similar to the proofs of Theorems 1 and 2, by making use of Lemmas 34 and 35 instead of Lemma 24. We leave the details to the reader. $\square$

4.4 Remarks on mth-level-valuations

In defining valuations, Kearns introduced a whole hierarchy of mth-level-valuations. However, it turns out that this is not necessary in the case for **S4** and **S5**. To this end, we first introduce another semantic consequence relation, and then prove its completeness with respect to the proof theory presented above. In the following, we will only deal with **S5** since the case for **S4** is exactly the same.

Definition 36. Let v be a function $v : \text{Form} \to \mathcal{V}$. Then, v is an **sS5**-*valuation* (simplified **S5**-valuation) iff v is mth-level-**S5**-valuation for every $m \in \{0, 1\}$.

Definition 37. A is an **sS5**-*tautology* ($\models_{\text{sS5}} A$) iff $v(A) = \mathbf{T}$ for all **sS5**-valuations v.

Remark 38. Compared to the earlier definition of **S5**-valuations, we only have two levels, instead of having the whole infinite hierarchy. Note here that the notion of an **sS5**-valuation and the notion of a 1st-level-**S5**-valuations are exactly the same.[7]

Now we prove the soundness and completeness. Let's begin with the soundness.

Proposition 39 (Soundness for **sS5**). If $\vdash_{\text{S5}} A$ then $\models_{\text{sS5}} A$.

Proof. It suffices to show that the axioms are **sS5**-tautologies, and the rules of inference preserve the truth value $\mathbf{T}$. For the axioms, we have by Proposition 33 that all axioms are legal-**S5**-tautologies. So, for all **sS5**-valuations, axioms are assigned the value $\mathbf{T}$. Thus it remains to check the validity of the rules.

For (MP), assume that A and $A \to B$ are **sS5**-tautologies. Then we have that $v(A) = v(A \to B) = \mathbf{T}$ for all **sS5**-valuations which is also a legal-**S5**-valuation, and by the truth table for $\to$, we obtain that $v(B) = \mathbf{T}$, i.e. B is a **sS5**-tautology, as desired. Similarly, for (RN), we assume that A is a **sS5**-tautology, and it follows that $v(A) = \mathbf{T}$ for all **sS5**-valuations. Therefore, by the truth table for $\square$, we obtain that $v(\square A) = \mathbf{T}$ for all **sS5**-valuations, i.e. $\square A$ is a **sS5**-tautology, as desired. This completes the proof. □

Remark 40. The key here is that once formulas take the value $\mathbf{T}$, then the boxed formulas also take the value $\mathbf{T}$. This is the feature enjoyed by **S4** and **S5**, but not by **T**.

For the completeness, we do not have to add anything new.

Theorem 4 (Completeness for **sS5**). If $\models_{\text{sS5}} A$ then $\vdash_{\text{S5}} A$.

---

[7] We would like to thank an anonymous referee for the suggestion to clarify this point.

Proof. We prove the contrapositive. Suppose that $\not\vdash_{\mathbf{S5}} A$. Then by Lemma 22, we have an A-saturated set Σ_0 such that $\Sigma_0 \not\vdash A$. In view of Lemma 35, v_{Σ_0} is a legal-$\mathbf{S5}$-valuation, and by Lemma 27, this v_{Σ_0} is also an $\mathbf{sS5}$-valuation. (Here, we only need a special case of Lemma 27 in which $m = 1$.) And since we have $v_{\Sigma_0}(A) \notin \mathcal{D}$, it is also the case that $v_{\Sigma_0}(A) \neq \mathbf{T}$ (since $v_{\Sigma_0}(A) = \mathbf{T}$ implies that $v_{\Sigma_0}(A) \in \mathcal{D}$) and thus we have $\not\models_{\mathbf{sS5}} A$, as desired. □

Remark 41. We have established that we do not need the whole hierarchy of valuations for **S4** and **S5**. One of the obvious properties of **S4** and **S5** is that there are only finitely many iterated modalities, and this is not the case in other modal logics we handle in this paper. There might be a deeper relation between the iterated modalities and the 'height' of the hierarchy, but we will leave this topic for further investigation.

Remark 42. Even though we were able to simplify the 'height' of the mth-level-valuations for the normal modal logics **S4** and **S5** from possibly infinite levels to only two levels, this does not mean that finding models or giving counter-models for formulas becomes an easy task. On the contrary, when calculating models (or counter models) of certain formulas one never uses full valuations but only partial valuations, i.e., only those valuations which are relevant for a certain sentence.[8] In Remark 10, we dropped a hint that Kearns' semantics lacks analyticity in languages with defined operators. But, this is even an issue in the language used in this paper. Take for instance the following **S4**-valid formula, $\Box(p \to q) \to \Box(\Box p \to \Box q)$. At first sight, it seems to be the case that this formula is not valid in Kearns' semantics by considering the following partial valuation:

$\Box$	$(p$	$\to$	$q)$	$\to$	$\Box$	$(\Box$	p	$\to$	$\Box$	$q)$
T	t	T	t	F	F	f	t	t	f	t

But, as pointed out in Remark 10, we have to take into account the valuations of *all* relevant legal-**S4**-tautologies. Then the problematic valuation in the given example is ruled out by the mth-level-**S4**-valuations of the axiom (LK). However, even though we can prove the existence of mth-level-valuations for every tautology (cf. Lemma 27 and Theorem 2), it is not obvious that we only need to check finitely many relevant formulas for any given formula.

This observation imposes a threat to those who want to establish this semantics as an alternative semantics for normal modal logics and their relatives, such as intuitionistic logic, in the context of computer science. Nevertheless, we are inclined to say, at least for the moment, that we are not concerned with decision procedures.

---

[8] We would like to thank an anonymous referee for pointing this out.

Our interest in this semantics, despite this paper being technical, is a philosophical one, and thus we leave this interesting topic for future investigation.

5 Generalizing the result of Kearns

One natural question regarding Kearns' result is to ask how we can capture other normal modal logics such as **K** with the same style. To the best of the authors' knowledge, this has not been addressed elsewhere, and so we will here give an answer to this. First, we deal with **K**, and then move on to **KD** and **KTB**.

5.1 Modal logic K

We first introduce the Nmatrix for the modal logic **K** which requires eight truth values.

Definition 43. A **K**-*Nmatrix* for $\mathcal{L}$ is a tuple $M = \langle \mathcal{V}, \mathcal{D}, \mathcal{O} \rangle$, where:

(a) $\mathcal{V} = \{\mathbf{T}, \mathbf{t_1}, \mathbf{t}, \mathbf{t_2}, \mathbf{f_2}, \mathbf{f}, \mathbf{f_1}, \mathbf{F}\}$,
(b) $\mathcal{D} = \{\mathbf{T}, \mathbf{t_1}, \mathbf{t}, \mathbf{t_2}\}$,
(c) For every n-ary connective $*$ of $\mathcal{L}$, $\mathcal{O}$ includes a corresponding n-ary function $\tilde{*}$ from $\mathcal{V}^n$ to $2^{\mathcal{V}} \setminus \{\emptyset\}$ as follows (we omit the brackets for sets):

A	$\neg A$	$\Box A$	$\Diamond A$		$A \dot{\rightarrow} B$	$\mathbf{T}$	$\mathbf{t_1}$	$\mathbf{t}$	$\mathbf{t_2}$	$\mathbf{f_2}$	$\mathbf{f}$	$\mathbf{f_1}$	$\mathbf{F}$
$\mathbf{T}$	$\mathbf{F}$	$\mathcal{D}$	$\mathcal{D}$		$\mathbf{T}$	$\mathbf{T}$	$\mathbf{t_1}$	$\mathbf{t}$	$\mathbf{t_2}$	$\mathbf{f_2}$	$\mathbf{f}$	$\mathbf{f_1}$	$\mathbf{F}$
$\mathbf{t_1}$	$\mathbf{f_1}$	$\mathcal{D}$	$\mathcal{F}$		$\mathbf{t_1}$	$\mathbf{T}$	$\mathbf{t_1}$	$\mathbf{t}$	$\mathbf{t_2}$	$\mathbf{f_2}$	$\mathbf{f}$	$\mathbf{f_1}$	$\mathbf{F}$
$\mathbf{t}$	$\mathbf{f}$	$\mathcal{F}$	$\mathcal{D}$		$\mathbf{t}$	$\mathbf{T}$	$\mathbf{T}$	$\mathbf{T}, \mathbf{t}$	$\mathbf{T}, \mathbf{t}$	$\mathbf{f_2}$	$\mathbf{f_2}, \mathbf{f}$	$\mathbf{f_2}$	$\mathbf{f_2}, \mathbf{f}$
$\mathbf{t_2}$	$\mathbf{f_2}$	$\mathcal{F}$	$\mathcal{F}$		$\mathbf{t_2}$	$\mathbf{T}$	$\mathbf{T}$	$\mathbf{T}, \mathbf{t}$	$\mathbf{T}, \mathbf{t}$	$\mathbf{f_2}$	$\mathbf{f_2}, \mathbf{f}$	$\mathbf{f_2}$	$\mathbf{f_2}, \mathbf{f}$
$\mathbf{f_2}$	$\mathbf{t_2}$	$\mathcal{D}$	$\mathcal{D}$		$\mathbf{f_2}$	$\mathbf{T}$	$\mathbf{t_1}$	$\mathbf{t}$	$\mathbf{t_2}$	$\mathbf{T}$	$\mathbf{t}$	$\mathbf{t_1}$	$\mathbf{t_2}$
$\mathbf{f}$	$\mathbf{t}$	$\mathcal{F}$	$\mathcal{D}$		$\mathbf{f}$	$\mathbf{T}$	$\mathbf{T}$	$\mathbf{T}, \mathbf{t}$	$\mathbf{T}, \mathbf{t}$	$\mathbf{T}$	$\mathbf{T}, \mathbf{t}$	$\mathbf{T}$	$\mathbf{T}, \mathbf{t}$
$\mathbf{f_1}$	$\mathbf{t_1}$	$\mathcal{D}$	$\mathcal{F}$		$\mathbf{f_1}$	$\mathbf{T}$	$\mathbf{t_1}$	$\mathbf{t}$	$\mathbf{t_2}$	$\mathbf{T}$	$\mathbf{t}$	$\mathbf{t_1}$	$\mathbf{t_2}$
$\mathbf{F}$	$\mathbf{T}$	$\mathcal{F}$	$\mathcal{F}$		$\mathbf{F}$	$\mathbf{T}$	$\mathbf{T}$	$\mathbf{T}, \mathbf{t}$	$\mathbf{T}, \mathbf{t}$	$\mathbf{T}$	$\mathbf{T}, \mathbf{t}$	$\mathbf{T}$	$\mathbf{T}, \mathbf{t}$

where $\mathcal{F} = \{\mathbf{f_2}, \mathbf{f}, \mathbf{f_1}, \mathbf{F}\}$.

A *legal-***K***-valuation* in a **K**-Nmatrix M is a function $v : \mathsf{Form} \to \mathcal{V}$ that satisfies the following condition for every n-ary connective $*$ of $\mathcal{L}$ and $A_1, \ldots, A_n \in \mathsf{Form}$: $v(*(A_1, \ldots, A_n)) \in \tilde{*}(v(A_1), \ldots, v(A_n))$.

Definition 44. A is a *legal-***K***-consequence* of Γ ($\Gamma \models_{\text{legal-K}} A$) iff for all legal valuations v, if $v(B) \in \mathcal{D}$ for all $B \in \Gamma$ then $v(A) \in \mathcal{D}$. In particular, A is a *legal-***K***-tautology* iff $v(A) \in \mathcal{D}$ for all legal-**K**-valuations v.

Definition 45. Let v be a function $v : \mathsf{Form} \to \mathcal{V}$. Then,

- v is a *0th-level-***K***-valuation* if v is a legal-**K**-valuation.

- v is a $m+1$st-level-**K**-valuation iff v is an mth-level-**K**-valuation and for every sentence A, $v(A) \in \{\mathbf{T}, \mathbf{t_1}\}$ holds if $v'(A) \in \mathcal{D}$ for all mth-level-**K**-valuations v'.

Based on these, we define v to be a **K**-*valuation* iff v is mth-level-**K**-valuation for every $m \geq 0$.

Definition 46. A is a **K**-*tautology* ($\models_\mathbf{K} A$) iff $v(A) \in \{\mathbf{T}, \mathbf{t_1}\}$ for all **K**-valuations v.

Remark 47. Note that we have the condition $v(A) \in \{\mathbf{T}, \mathbf{t_1}\}$ for **K**-valuations, not only $v(A) = \mathbf{T}$.

We now turn to the proof theory. Again, we introduce two systems, one being the well-known system **K**, and the other is a subsystem of **K**.

Definition 48. First, the system **K** is obtained by eliminating the axiom (LT) from the system **T**. Second, we define a subsystem of **K**, referred to as $\mathbf{K}^-$, which is obtained by eliminating (RN) and adding the following formulas:

$\Diamond \neg A \rightarrow \neg \Box A$	(LM3)	$\Box \neg (A \rightarrow B) \rightarrow \Box A$	(LK3)
$\neg \Box A \rightarrow \Diamond \neg A$	(LM4)	$\Box \neg (A \rightarrow B) \rightarrow \Box \neg B$	(LK4)
$\Diamond (A \rightarrow B) \rightarrow (\Box A \rightarrow \Diamond B)$	(LK2)	$\Diamond \neg (A \rightarrow B) \rightarrow \Diamond \neg B$	(LK6)

The consequence relations are defined as in Definition 11.

We now turn to prove the soundness and completeness. First, we deal with the soundness result.

Proposition 49 (Soundness for $\mathbf{K}^-$). *If* $\Gamma \vdash_{\mathbf{K}^-} A$ *then* $\Gamma \models_{\text{legal-K}} A$.

Proof. It suffices to check that all axioms are legal-**K**-tautologies, and that the rule of inference (MP) preserves the designated value. $\square$

The following is the analogue of Lemma 15.

Lemma 50. *Assume that* $\vdash_\mathbf{K} A$ *and that the length of the proof for A is m. Then, for all mth-level-**K**-valuations v, $v(A) \in \mathcal{D}$.*

Proof. Similar to the proof of Lemma 15. Details are left to the reader. $\square$

Once we have the lemma, the soundness follows immediately.

Proposition 51 (Soundness for **K**). *If* $\vdash_\mathbf{K} A$ *then* $\models_\mathbf{K} A$.

Proof. Let the length of the proof for A be m. Then, by the above lemma, A is true for every mth-level-**K**-valuation. Since **K**-valuations are also mth-level-**K**-valuations, we obtain that A is true for all **K**-valuations, as desired. □

Now we turn to prove the completeness.

Lemma 52. Let Σ be a **K**-relatively maximal set, and define a function v_Σ from Form to $\mathcal{V}$ as follows:

$$v_\Sigma(B) := \begin{cases} \mathbf{T} & \text{if } \Sigma \vdash \Box B \text{ and } \Sigma \vdash B \text{ and } \Sigma \vdash \Diamond B \\ \mathbf{t_1} & \text{if } \Sigma \vdash \Box B \text{ and } \Sigma \vdash B \text{ and } \Sigma \nvdash \Diamond B \\ \mathbf{t} & \text{if } \Sigma \nvdash \Box B \text{ and } \Sigma \vdash B \text{ and } \Sigma \vdash \Diamond B \\ \mathbf{t_2} & \text{if } \Sigma \nvdash \Box B \text{ and } \Sigma \vdash B \text{ and } \Sigma \nvdash \Diamond B \\ \mathbf{f_2} & \text{if } \Sigma \vdash \Box B \text{ and } \Sigma \nvdash B \text{ and } \Sigma \vdash \Diamond B \\ \mathbf{f} & \text{if } \Sigma \nvdash \Box B \text{ and } \Sigma \nvdash B \text{ and } \Sigma \vdash \Diamond B \\ \mathbf{f_1} & \text{if } \Sigma \vdash \Box B \text{ and } \Sigma \nvdash B \text{ and } \Sigma \nvdash \Diamond B \\ \mathbf{F} & \text{if } \Sigma \nvdash \Box B \text{ and } \Sigma \nvdash B \text{ and } \Sigma \nvdash \Diamond B \end{cases}$$

Then, v_Σ is a well-defined legal-**K**-valuation.

Proof. The details are spelled out in the appendix. □

Remark 53. The reason we need eight values is partially explained in the above Lemma 52. Indeed, unlike the extensions of the modal logic **T**, necessity and possibility behave in an "independent" manner, in the sense that the former does not necessarily imply the latter, and thus we need more combinations of modalities. Also note that if we eliminate the values $\mathbf{t_1}, \mathbf{t_2}, \mathbf{f_1}$ and $\mathbf{f_2}$, then we obtain the definition of the legal-**T**-valuation (cf. Definition 23).

Theorem 5 (Completeness for $\mathbf{K^-}$). *If* $\Gamma \models_{\text{legal-}\mathbf{K}} A$ *then* $\Gamma \vdash_{\mathbf{K^-}} A$.

Proof. Similar to the proof of Theorem 1, by Lemma 52 instead of Lemma 24. We leave the details to the reader. □

Lemma 54. Let Γ be a **K**-relatively maximal set. If v_Γ is a legal-**K**-valuation, then v_Γ is also an mth-level-**K**-valuation for every $m \geq 1$, and thus a **K**-valuation.

Proof. Similar to the proof of Lemma 27. We leave the details to the reader. □

Theorem 6 (Completeness for **K**). *If* $\models_\mathbf{K} A$ *then* $\vdash_\mathbf{K} A$.

Proof. Similar to the proof of Theorem 2, by Lemmas 52 and 54 instead of Lemmas 24 and 27 respectively. We leave the details to the reader. □

5.2 Modal logic KD

The semantics will be based on a six-valued Nmatrix. More specifically, the truth values t_1, f_1 will be eliminated from the **K**-Nmatrix in order to obtain the Nmatrix for **KD**.

Definition 55. A **D**-*Nmatrix* for $\mathcal{L}$ is a tuple $M = \langle \mathcal{V}, \mathcal{D}, \mathcal{O} \rangle$, where:

(a) $\mathcal{V} = \{\mathbf{T}, \mathbf{t}, \mathbf{t_2}, \mathbf{f_2}, \mathbf{f}, \mathbf{F}\}$,
(b) $\mathcal{D} = \{\mathbf{T}, \mathbf{t}, \mathbf{t_2}\}$,
(c) For every n-ary connective $*$ of $\mathcal{L}$, $\mathcal{O}$ includes a corresponding n-ary function $\tilde{*}$ from $\mathcal{V}^n$ to $2^{\mathcal{V}} \setminus \{\emptyset\}$ as follows (we omit the brackets for sets):

A	$\tilde{\neg}A$	$\tilde{\Box}A$	$\tilde{\Diamond}A$
$\mathbf{T}$	$\mathbf{F}$	$\mathcal{D}$	$\mathcal{D}$
$\mathbf{t}$	$\mathbf{f}$	$\mathcal{F}$	$\mathcal{D}$
$\mathbf{t_2}$	$\mathbf{f_2}$	$\mathcal{F}$	$\mathcal{F}$
$\mathbf{f_2}$	$\mathbf{t_2}$	$\mathcal{D}$	$\mathcal{D}$
$\mathbf{f}$	$\mathbf{t}$	$\mathcal{F}$	$\mathcal{D}$
$\mathbf{F}$	$\mathbf{T}$	$\mathcal{F}$	$\mathcal{F}$

$A\tilde{\rightarrow}B$	$\mathbf{T}$	$\mathbf{t}$	$\mathbf{t_2}$	$\mathbf{f_2}$	$\mathbf{f}$	$\mathbf{F}$
$\mathbf{T}$	$\mathbf{T}$	$\mathbf{t}$	$\mathbf{t_2}$	$\mathbf{f_2}$	$\mathbf{f}$	$\mathbf{F}$
$\mathbf{t}$	$\mathbf{T}$	$\mathbf{T},\mathbf{t}$	$\mathbf{T},\mathbf{t}$	$\mathbf{f_2}$	$\mathbf{f_2},\mathbf{f}$	$\mathbf{f_2},\mathbf{f}$
$\mathbf{t_2}$	$\mathbf{T}$	$\mathbf{T},\mathbf{t}$	$\mathbf{T},\mathbf{t}$	$\mathbf{f_2}$	$\mathbf{f_2},\mathbf{f}$	$\mathbf{f_2},\mathbf{f}$
$\mathbf{f_2}$	$\mathbf{T}$	$\mathbf{t}$	$\mathbf{t_2}$	$\mathbf{T}$	$\mathbf{t}$	$\mathbf{t_2}$
$\mathbf{f}$	$\mathbf{T}$	$\mathbf{T},\mathbf{t}$	$\mathbf{T},\mathbf{t}$	$\mathbf{T}$	$\mathbf{T},\mathbf{t}$	$\mathbf{T},\mathbf{t}$
$\mathbf{F}$	$\mathbf{T}$	$\mathbf{T},\mathbf{t}$	$\mathbf{T},\mathbf{t}$	$\mathbf{T}$	$\mathbf{T},\mathbf{t}$	$\mathbf{T},\mathbf{t}$

where $\mathcal{F} = \{\mathbf{f_2}, \mathbf{f}, \mathbf{F}\}$.

A *legal-**D**-valuation* in a **D**-Nmatrix M is a function $v : \text{Form} \rightarrow \mathcal{V}$ that satisfies the following condition for every n-ary connective $*$ of $\mathcal{L}$ and $A_1, \ldots, A_n \in \text{Form}$: $v(*(A_1, \ldots, A_n)) \in \tilde{*}(v(A_1), \ldots, v(A_n))$.

Definition 56. A is a *legal-**D**-consequence of* Γ ($\Gamma \models_{\text{legal-D}} A$) iff for all legal-**D**-valuations v, if $v(B) \in \mathcal{D}$ for all $B \in \Gamma$ then $v(A) \in \mathcal{D}$. In particular, A is a *legal-**D**-tautology* iff $v(A) \in \mathcal{D}$ for all legal-**D**-valuations v.

Definition 57. Let v be a function $v : \text{Form} \rightarrow \mathcal{V}$. Then,

- v is a *0th-level-**D**-valuation* if v is a legal-**D**-valuation.
- v is a $m+1$*st-level-**D**-valuation* iff v is an mth-level-**D**-valuation and assigns value the $\mathbf{T}$ to every sentence A if $v'(A) \in \mathcal{D}$ for all mth-level-**D**-valuations v'.

Based on these, we define v to be a **D**-*valuation* iff v is mth-level-**D**-valuation for every $m \geq 0$.

Definition 58. A is a **D**-*tautology* ($\models_{\mathbf{D}} A$) iff $v(A) = \mathbf{T}$ for all **D**-valuations v.

Proof theory is obtained in an expected manner.

Definition 59. The systems **D** and **D**⁻ are obtained by adding the following formula to **K** and **K**⁻ respectively.
$$\Box A \to \Diamond A \tag{D1}$$

The consequence relations are defined as in Definition 11.

Remark 60. Note that the above **D**-Nmatrix can be refined by adding further axioms to **D**⁻. More specifically, we may replace the Nmatrix for $\to$ by the following Nmatrix, which can be found in [5, p.35], if we add axioms (LK1) and (LK5).

$A\tilde{\to}B$	T	t	t_2	f_2	f	F
T	T	t	t_2	f_2	f	F
t	T	T,t	t	f_2	f_2,f	f
t_2	T	T	T	f_2	f_2	f_2
f_2	T	t	t_2	T	t	t_2
f	T	T,t	t	T	T,t	t
F	T	T	T	T	T	T

However, we presented the above Nmatrix as it is obtained straightforwardly from our Nmatrix for **K** which is not presented elsewhere to the best of the authors' knowledge.

Now, we first consider the soundness.

Proposition 61 (Soundness for **D**⁻ and **D**). *If $\Gamma \vdash_{\mathbf{D}^-} A$ then $\Gamma \models_{\text{legal-}\mathbf{D}} A$ and if $\vdash_{\mathbf{D}} A$ then $\models_{\mathbf{D}} A$.*

Proof. For the former, it suffices to check that all axioms are legal-**D**-tautologies, and that the rule of inference (MP) preserves the designated values. For the latter, exactly as in Proposition 51. □

For the completeness proof, we need the following as before.

Lemma 62. *Let Σ be a **D**-relatively maximal set, and define a function v_Σ from Form to $\mathcal{V}$ as follows:*

$$v_0(B) := \begin{cases} \mathbf{T} & \text{if } \Sigma \vdash \Box B \text{ and } \Sigma \vdash B \text{ and } \Sigma \vdash \Diamond B \\ \mathbf{t} & \text{if } \Sigma \nvdash \Box B \text{ and } \Sigma \vdash B \text{ and } \Sigma \vdash \Diamond B \\ \mathbf{t_2} & \text{if } \Sigma \nvdash \Box B \text{ and } \Sigma \vdash B \text{ and } \Sigma \nvdash \Diamond B \\ \mathbf{f_2} & \text{if } \Sigma \vdash \Box B \text{ and } \Sigma \nvdash B \text{ and } \Sigma \vdash \Diamond B \\ \mathbf{f} & \text{if } \Sigma \nvdash \Box B \text{ and } \Sigma \nvdash B \text{ and } \Sigma \vdash \Diamond B \\ \mathbf{F} & \text{if } \Sigma \nvdash \Box B \text{ and } \Sigma \nvdash B \text{ and } \Sigma \nvdash \Diamond B \end{cases}$$

*Then, v_Σ is a well-defined legal-**D**-valuation.*

Proof. Note first that v_0 is well-defined in view of (D1). Then the desired result can be proved by induction on the number n of connectives.
(Base): For atomic formulas, just note that Σ is consistent which implies that $\Sigma \vdash A$ or $\Sigma \vdash \neg A$.
(Induction step): We split the cases based on the connectives. The details are exactly the same as in the case for **K** except that we eliminate the values $\mathbf{t_1}, \mathbf{f_1}$. □

Remark 63. The above lemma is easily obtained by eliminating the two values $\mathbf{t_1}$ and $\mathbf{f_1}$ in Lemma 52. Indeed, the conditions for the two values become not satisfiable in view of the additional axiom (D1).

Now we can prove the completeness.

Theorem 7 (Completeness for **D⁻** and **D**). If $\Gamma \models_{\text{legal-D}} A$ then $\Gamma \vdash_{\mathbf{D^-}} A$, and if $\models_{\mathbf{D}} A$ then $\vdash_{\mathbf{D}} A$.

Proof. Similar to the proofs of Theorems 1 and 6, by Lemma 62 instead of Lemma 24. We leave the details to the reader. □

5.3 Modal logic KTB

Again, the semantics will be based on a six-valued Nmatrix. Roughly speaking, we need to split the values **t** and **f** into two values respectively.

Definition 64. A **B**-*Nmatrix* for $\mathcal{L}$ is a tuple $M = \langle \mathcal{V}, \mathcal{D}, \mathcal{O} \rangle$, where:

(a) $\mathcal{V} = \{\mathbf{T}, \mathbf{t_3}, \mathbf{t_4}, \mathbf{f_4}, \mathbf{f_3}, \mathbf{F}\}$,
(b) $\mathcal{D} = \{\mathbf{T}, \mathbf{t_3}, \mathbf{t_4}\}$,
(c) For every n-ary connective $*$ of $\mathcal{L}$, $\mathcal{O}$ includes a corresponding n-ary function $\tilde{*}$ from $\mathcal{V}^n$ to $2^{\mathcal{V}} \setminus \{\emptyset\}$ as follows (we omit the brackets for sets):

A	$\neg A$	$\Box A$	$\Diamond A$	$A \tilde{\to} B$	**T**	$\mathbf{t_3}$	$\mathbf{t_4}$	$\mathbf{f_4}$	$\mathbf{f_3}$	**F**
T	**F**	$\mathcal{D}$	**T**	**T**	**T**	$\mathbf{t_3},\mathbf{t_4}$	$\mathbf{t_3},\mathbf{t_4}$	$\mathbf{f_4},\mathbf{f_3}$	$\mathbf{f_4},\mathbf{f_3}$	**F**
$\mathbf{t_3}$	$\mathbf{f_3}$	$\mathbf{f_4},\mathbf{f_3}$	**T**	$\mathbf{t_3}$	**T**	$\mathcal{D}$	$\mathcal{D}$	$\mathbf{f_3},\mathbf{f_4}$	$\mathbf{f_4},\mathbf{f_3}$	$\mathbf{f_4},\mathbf{f_3}$
$\mathbf{t_4}$	$\mathbf{f_4}$	**F**	**T**	$\mathbf{t_4}$	**T**	$\mathcal{D}$	$\mathcal{D}$	$\mathbf{f_3},\mathbf{f_4}$	$\mathbf{f_4},\mathbf{f_3}$	$\mathbf{f_4},\mathbf{f_3}$
$\mathbf{f_4}$	$\mathbf{t_4}$	**F**	**T**	$\mathbf{f_4}$	**T**	$\mathcal{D}$	$\mathcal{D}$	$\mathcal{D}$	$\mathcal{D}$	$\mathbf{t_3},\mathbf{t_4}$
$\mathbf{f_3}$	$\mathbf{t_3}$	**F**	$\mathbf{t_3},\mathbf{t_4}$	$\mathbf{f_3}$	**T**	$\mathcal{D}$	$\mathcal{D}$	$\mathcal{D}$	$\mathcal{D}$	$\mathbf{t_3},\mathbf{t_4}$
F	**T**	**F**	$\mathcal{F}$	**F**	**T**	**T**	**T**	**T**	**T**	**T**

where $\mathcal{F} = \{\mathbf{f_4}, \mathbf{f_3}, \mathbf{F}\}$.

A *legal*-**B**-*valuation* in a **B**-Nmatrix M is a function $v : \text{Form} \to \mathcal{V}$ that satisfies the following condition for every n-ary connective $*$ of $\mathcal{L}$ and $A_1, \ldots, A_n \in \text{Form}$: $v(*(A_1, \ldots, A_n)) \in \tilde{*}(v(A_1), \ldots, v(A_n))$.

Definition 65. A is a *legal-B-consequence of* Γ ($\Gamma \models_{\text{legal-B}} A$) iff for all legal-**B**-valuations v, if $v(B) \in \mathcal{D}$ for all $B \in \Gamma$ then $v(A) \in \mathcal{D}$. In particular, A is a *legal-B-tautology* iff $v(A) \in \mathcal{D}$ for all legal-**B**-valuations v.

Definition 66. Let v be a function $v : \text{Form} \to \mathcal{V}$. Then,

- v is a *0th-level-B-valuation* if v is a legal-**B**-valuation.
- v is a $m+1$*st-level-B-valuation* iff v is an mth-level-**B**-valuation and assigns the value **T** to every sentence A if $v'(A) \in \mathcal{D}$ for all mth-level-**B**-valuations v'.

Based on these, we define v to be a **B**-*valuation* iff v is an mth-level-**B**-valuation for every $m \geq 0$.

Definition 67. A is a **B**-*tautology* ($\models_{\mathbf{B}} A$) iff $v(A) = \mathbf{T}$ for all **B**-valuations v.

Now we introduce the proof theory.

Definition 68. The system **B** is obtained by adding (LB) to **T**. Moreover, the system $\mathbf{B}^-$ is obtained by adding the following formulas to $\mathbf{T}^-$:

$$A \to \Box \Diamond A \quad \text{(LB)} \qquad \neg \Diamond \Box A \to \Box \Diamond \neg A \quad \text{(B2)}$$
$$\Diamond \Box A \to A \quad \text{(MB)} \qquad \Diamond \Box \neg A \to \neg \Box \Diamond A \quad \text{(B3)}$$
$$\Box \Diamond \neg A \to \neg \Diamond \Box A \quad \text{(B1)} \qquad \neg \Box \Diamond A \to \Diamond \Box \neg A \quad \text{(B4)}$$

The consequence relations are defined as in Definition 11.

We will use the following in the completeness proof.

Proposition 69. The following formulas are provable in $\mathbf{B}^-$:

$$\Diamond \Box A \to \Diamond \Box \neg \neg A \quad (10) \qquad \Diamond \Box \neg \neg A \to \Diamond \Box A \quad (11) \qquad \neg A \to \Box \neg \Box A \quad (12)$$

Proof. (10) is provable by (B1) and (B4), and (11) is provable by (B3) and (B2). Finally, (12) is provable by making use of (MB) and (LM2). □

We now turn to prove the soundness and completeness. As expected, soundness is straightforward.

Proposition 70 (Soundness for $\mathbf{B}^-$ and **B**). *If* $\Gamma \vdash_{\mathbf{B}^-} A$ *then* $\Gamma \models_{\text{legal-B}} A$ *and if* $\vdash_{\mathbf{B}} A$ *then* $\models_{\mathbf{B}} A$.

Proof. For the former, it suffices to check that all axioms are legal-**B**-tautologies, and that the rule of inference (MP) preserves the designated values. For the latter, the proof runs exactly as in Proposition 51. □

For the completeness proof, we need the following as before.

Lemma 71. Let Σ be a **B**-relatively maximal set, and define a function v_Σ from Form to $\mathcal{V}$ as follows:

$$v_0(B) := \begin{cases} \mathbf{T} & \text{if } \Sigma \vdash \Box B \\ \mathbf{t_3} & \text{if } \Sigma \not\vdash \Box B \text{ and } \Sigma \vdash \Diamond\Box B \text{ and } \Sigma \vdash B \\ \mathbf{t_4} & \text{if } \Sigma \not\vdash \Box B \text{ and } \Sigma \not\vdash \Diamond\Box B \text{ and } \Sigma \vdash B \\ \mathbf{f_4} & \text{if } \Sigma \not\vdash \Box\neg B \text{ and } \Sigma \not\vdash \Diamond\Box\neg B \text{ and } \Sigma \vdash \neg B \\ \mathbf{f_3} & \text{if } \Sigma \not\vdash \Box\neg B \text{ and } \Sigma \vdash \Diamond\Box\neg B \text{ and } \Sigma \vdash \neg B \\ \mathbf{F} & \text{if } \Sigma \vdash \Box\neg B \end{cases}$$

Then, v_Σ is a well-defined legal-**B**-valuation.

Proof. The details are spelled out in the appendix. □

Remark 72. If we take the disjunction of the conditions for $\mathbf{t_3}$ and $\mathbf{t_4}$ ($\mathbf{f_3}$ and $\mathbf{f_4}$), then we obtain the condition for the value $\mathbf{t}$ ($\mathbf{f}$).

Now we can prove the completeness.

Theorem 8 (Completeness for $\mathbf{B}^-$ and $\mathbf{B}$). *If $\Gamma \models_{\text{legal-}\mathbf{B}} A$ then $\Gamma \vdash_{\mathbf{B}^-} A$, and if $\models_\mathbf{B} A$ then $\vdash_\mathbf{B} A$.*

Proof. Similar to the proofs of Theorems 1 and 6, by Lemma 71 instead of Lemma 24. We leave the details to the reader. □

6 Conclusion

The present paper was focusing on the technical aspects of Kearns' result. We clarified his original approach and put it in the context of non-deterministic semantics. Some of the results we presented have been already reported independently in [5]. The results not included in [5] that deserve special attention are as follows. First, we simplified the semantics of Kearns for **S4** and **S5** (cf. §4.4). Second, we introduced Kearns' style semantics for **K** and proved its soundness and completeness with respect to the Hilbert-style proof theory (cf. §5.1). Finally, we pointed out an error in the results of Ivlev (cf. §3.3).

In view of the result that Kripke's semantics and Kearns' semantics are equivalent, we now leave it to the reader to decide if the latter is challenging the former on an ontological level. We understand the point of Kearns in claiming that his semantics is free of possible worlds, and if one shares the view of Kearns, then his semantics certainly has an advantage. However, in order to justify Kearns' approach philosophically for a wider audience, a lot of additional work still has to be done. Kearns himself claimed that his approach is simplier than Kripke's semantics for normal modal logics since it has dispensed with possible worlds and, therefore, is preferable (cf. [11, p.86]). Kearns did not believe in possible worlds. But the story goes further, since now, e.g. for **S5**, we have to justify at least three issues, namely four different truth-values (instead of two truth-values), non-deterministic matrices for the non-modal connectives (instead of a deterministic matrices) and at least two levels of valuations (instead of one level). If we compare this to the fact that we only need to justify possible worlds for Kripkean semantics, the claim of Kearns seems to be a bit too bold.

We just note in passing by that we can consider a two-valued, deterministic and non-hierarchic semantics which overcomes all the difficulties of Kearns' semantics and also does not involve possible worlds. The resulting system will be something very close to the proposal of Jean-Yves Béziau in [3], being a strict extension of modal logic **S5**. Note that the presentation in [3] deploys four-valued semantics, but this can be easily turned into a two-valued semantics in the manner of Michael Dunn's relational semantics. The point to be emphasized here is that despite the argument of Josep Maria Font and Petr Hájek in [8] against Jan Łukasiewicz, the many-valued approach to modality deserves further attention. We shall keep the details for another occasion.

We finish this paper by pointing out three possible directions for future research. First, considering the tight connection between non-deterministic semantics and sequent calculi established by Avron and his collaborators, it is interesting to examine proof-theoretical implications of Kearns' semantics. Due to the hierarchical aspect of Kearns' semantics, it is not straightforward to obtain a new perspective on the proof theory of modal logics, but this seems to be an interesting topic.

Second, one of the virtues of Kripkean semantics is the correspondence between axioms and the accessibility relations of the Kripke frame and it is very natural to ask if something similar holds in Kearns' non-deterministic semantics. Note that many of the non-deterministic semantics given by Avron and his colleagues are modular. But a glance at the semantics for the systems **K, KD, KTB, T, S4** and **S5** introduced here reveals that if there is a correspondence it is not a simple one, since we need to change not only the truth tables of $\Box$ and $\Diamond$ but also the 'height' of the mth level valuations, the set of truth-values and the set of designated values as well.

Finally, the usual possible worlds semantics provide decision procedures for modal logics. However, as pointed out in Remark 42, we cannot guarantee that this will also be the case for Kearn's semantics. We will leave this interesting topic for future research.

In conclusion, we hope to have shown the richness of Kearns' semantics which deserves further attention towards shedding some new light on modal logics.

References

[1] Arnon Avron and Iddo Lev. Non-Deterministic Multiple-valued Structures. *Journal of Logic and Computation*, 15(3):241–261, 2005.

[2] Arnon Avron and Anna Zamansky. Non-deterministic semantics for logical systems. In *Handbook of Philosophical Logic*, volume 16, pages 227–304. Springer, 2011.

[3] Jean-Yves Béziau. A new four-valued approach to modal logic. *Logique et Analyse*, 54(213):109–121, 2011.

[4] Brian F. Chellas. *Modal Logic*. Cambridge University Press, 1980.

[5] Marcelo E. Coniglio, Luis Fariñas del Cerro, and Newton Peron. Finite non-deterministic semantics for some modal systems. *Journal of Applied Non-Classical Logics*, 25(1):20–45, 2015.

[6] Luis Fariñas del Cerro, Marcelo E. Coniglio, and Newton Peron. Non-deterministic matrices for modal logic. in p.47 of Book of abstracts for 17th Brazilian Logic Conference, 2014. Available at: http://www.uff.br/ebl/EBL_2014_book_of_abstracts.pdf.

[7] Thomas Macaulay Ferguson. On non-deterministic quantification. *Logica Universalis*, 8:165–191, 2014.

[8] Josep Maria Font and Petr Hájek. On Łukasiewicz's Four-Valued Modal Logic. *Studia Logica*, 70(2):157–182, 2002.

[9] Yuri. V. Ivlev. A semantics for modal calculi. *Bulletin of the Section of Logic*, 17(3/4):114–121, 1988.

[10] Yuri. V. Ivlev. *Modal logic. (in Russian)*. Moskva: Moskovskij Gosudarstvennyj Universitet, 1991.

[11] John Kearns. Modal Semantics without Possible Worlds. *Journal of Symbolic Logic*, 46(1):77–86, 1981.

[12] John Kearns. Leśniewski's strategy and modal logic. *Notre Dame Journal of Formal Logic*, 30(2):77–86, 1989.

Appendix

In this appendix, we spell out the details of the proofs for Lemmas 52 and 71.

Proof for Lemma 52

Note first that v_0 is well-defined. The desired result can be proved by induction on the number n of connectives.

(**Base**): For atomic formulas, just note that Σ is consistent which implies that $\Sigma \vdash A$ or $\Sigma \vdash \neg A$.

(**Induction step**): We split the cases based on the connectives.

Case 1. If $B = \neg C$, then we have the following eight cases.

cases	$v_\Sigma(C)$	condition for C	$v_\Sigma(B)$	condition for B i.e. $\neg C$
(i)	**T**	$\Sigma \vdash \Box C$ & $\Sigma \vdash C$ & $\Sigma \vdash \Diamond C$	**F**	$\Sigma \nvdash \Box \neg C$ & $\Sigma \nvdash \neg C$ & $\Sigma \nvdash \Diamond \neg C$
(ii)	t_1	$\Sigma \vdash \Box C$ & $\Sigma \vdash C$ & $\Sigma \nvdash \Diamond C$	f_1	$\Sigma \vdash \Box \neg C$ & $\Sigma \nvdash \neg C$ & $\Sigma \nvdash \Diamond \neg C$
(iii)	t	$\Sigma \nvdash \Box C$ & $\Sigma \vdash C$ & $\Sigma \vdash \Diamond C$	f	$\Sigma \nvdash \Box \neg C$ & $\Sigma \nvdash \neg C$ & $\Sigma \vdash \Diamond \neg C$
(iv)	t_2	$\Sigma \nvdash \Box C$ & $\Sigma \vdash C$ & $\Sigma \nvdash \Diamond C$	f_2	$\Sigma \vdash \Box \neg C$ & $\Sigma \nvdash \neg C$ & $\Sigma \vdash \Diamond \neg C$
(v)	f_2	$\Sigma \vdash \Box C$ & $\Sigma \nvdash C$ & $\Sigma \vdash \Diamond C$	t_2	$\Sigma \nvdash \Box \neg C$ & $\Sigma \vdash \neg C$ & $\Sigma \nvdash \Diamond \neg C$
(vi)	f	$\Sigma \nvdash \Box C$ & $\Sigma \nvdash C$ & $\Sigma \vdash \Diamond C$	t	$\Sigma \nvdash \Box \neg C$ & $\Sigma \vdash \neg C$ & $\Sigma \vdash \Diamond \neg C$
(vii)	f_1	$\Sigma \vdash \Box C$ & $\Sigma \nvdash C$ & $\Sigma \nvdash \Diamond C$	t_1	$\Sigma \vdash \Box \neg C$ & $\Sigma \vdash \neg C$ & $\Sigma \nvdash \Diamond \neg C$
(viii)	**F**	$\Sigma \nvdash \Box C$ & $\Sigma \nvdash C$ & $\Sigma \nvdash \Diamond C$	**F**	$\Sigma \vdash \Box \neg C$ & $\Sigma \vdash \neg C$ & $\Sigma \vdash \Diamond \neg C$

By induction hypothesis, we have the conditions for C, and it is easy to see that the conditions for B i.e. $\neg C$ are provable.

Case 2. If $B = \Box C$, then we have the following eight cases.

cases	$v_\Sigma(C)$	condition for C	$v_\Sigma(B)$	condition for B i.e. $\Box C$
(i)	**T**	$\Sigma \vdash \Box C$ and $\Sigma \vdash C$ and $\Sigma \vdash \Diamond C$	$\mathbf{T}, t_1, t, t_2$	$\Sigma \vdash \Box C$
(ii)	t_1	$\Sigma \vdash \Box C$ and $\Sigma \vdash C$ and $\Sigma \nvdash \Diamond C$	$\mathbf{T}, t_1, t, t_2$	$\Sigma \vdash \Box C$
(iii)	t	$\Sigma \nvdash \Box C$ and $\Sigma \vdash C$ and $\Sigma \vdash \Diamond C$	$\mathbf{F}, f_1, f, f_2$	$\Sigma \nvdash \Box C$
(iv)	t_2	$\Sigma \nvdash \Box C$ and $\Sigma \vdash C$ and $\Sigma \nvdash \Diamond C$	$\mathbf{F}, f_1, f, f_2$	$\Sigma \nvdash \Box C$
(v)	f_2	$\Sigma \vdash \Box C$ and $\Sigma \nvdash C$ and $\Sigma \vdash \Diamond C$	$\mathbf{T}, t_1, t, t_2$	$\Sigma \vdash \Box C$
(vi)	f	$\Sigma \nvdash \Box C$ and $\Sigma \nvdash C$ and $\Sigma \vdash \Diamond C$	$\mathbf{F}, f_1, f, f_2$	$\Sigma \nvdash \Box C$
(vii)	f_1	$\Sigma \vdash \Box C$ and $\Sigma \nvdash C$ and $\Sigma \nvdash \Diamond C$	$\mathbf{T}, t_1, t, t_2$	$\Sigma \vdash \Box C$
(viii)	**F**	$\Sigma \nvdash \Box C$ and $\Sigma \nvdash C$ and $\Sigma \nvdash \Diamond C$	$\mathbf{F}, f_1, f, f_2$	$\Sigma \nvdash \Box C$

By induction hypothesis, we have the conditions for C, and we can see that the conditions for B i.e. $\Box C$ are provable.

Case 3. If $B = \Diamond C$, then we have the following eight cases.

cases	$v_\Sigma(C)$	condition for C	$v_\Sigma(B)$	condition for B i.e. $\Box C$
(i)	**T**	$\Sigma \vdash \Box C$ and $\Sigma \vdash C$ and $\Sigma \vdash \Diamond C$	$\mathbf{T}, t_1, t, t_2$	$\Sigma \vdash \Diamond C$
(ii)	t_1	$\Sigma \vdash \Box C$ and $\Sigma \vdash C$ and $\Sigma \nvdash \Diamond C$	$\mathbf{F}, f_1, f, f_2$	$\Sigma \nvdash \Diamond C$
(iii)	t	$\Sigma \nvdash \Box C$ and $\Sigma \vdash C$ and $\Sigma \vdash \Diamond C$	$\mathbf{T}, t_1, t, t_2$	$\Sigma \vdash \Diamond C$
(iv)	t_2	$\Sigma \nvdash \Box C$ and $\Sigma \vdash C$ and $\Sigma \nvdash \Diamond C$	$\mathbf{F}, f_1, f, f_2$	$\Sigma \nvdash \Diamond C$
(v)	f_2	$\Sigma \vdash \Box C$ and $\Sigma \nvdash C$ and $\Sigma \vdash \Diamond C$	$\mathbf{T}, t_1, t, t_2$	$\Sigma \vdash \Diamond C$
(vi)	f	$\Sigma \nvdash \Box C$ and $\Sigma \nvdash C$ and $\Sigma \vdash \Diamond C$	$\mathbf{T}, t_1, t, t_2$	$\Sigma \vdash \Diamond C$
(vii)	f_1	$\Sigma \vdash \Box C$ and $\Sigma \nvdash C$ and $\Sigma \nvdash \Diamond C$	$\mathbf{F}, f_1, f, f_2$	$\Sigma \nvdash \Diamond C$
(viii)	**F**	$\Sigma \nvdash \Box C$ and $\Sigma \nvdash C$ and $\Sigma \nvdash \Diamond C$	$\mathbf{F}, f_1, f, f_2$	$\Sigma \nvdash \Diamond C$

By induction hypothesis, we have the conditions for C, and we can see that the conditions for B i.e. $\Box C$ are provable.

Case 4. If $B = C \rightarrow D$, then we have the following 21 cases.

cases	$v_\Sigma(C)$	condition for C	$v_\Sigma(D)$	condition for D	$v_\Sigma(B)$	condition for B i.e. $C{\to}D$
(1)	any	—	T	$\Sigma \vdash \Box D$ & $\Sigma \vdash D$ & $\Sigma \vdash \Diamond D$	T	$\Sigma \vdash \Box B$ & $\Sigma \vdash B$ & $\Sigma \vdash \Diamond B$
(2)	T, t_1, f_2, f_1	$\Sigma \vdash \Box C$	t_1	$\Sigma \vdash \Box D$ & $\Sigma \vdash D$ & $\Sigma \not\vdash \Diamond D$	t_1	$\Sigma \vdash \Box B$ & $\Sigma \vdash B$ & $\Sigma \not\vdash \Diamond B$
(3)	t, t_2, f, F	$\Sigma \not\vdash \Box C$	t_1	$\Sigma \vdash \Box D$ & $\Sigma \vdash D$ & $\Sigma \not\vdash \Diamond D$	T	$\Sigma \vdash \Box B$ & $\Sigma \vdash B$ & $\Sigma \vdash \Diamond B$
(4)	T, t_1, f_2, f_1	$\Sigma \vdash \Box C$	t	$\Sigma \not\vdash \Box D$ & $\Sigma \vdash D$ & $\Sigma \vdash \Diamond D$	t	$\Sigma \not\vdash \Box B$ & $\Sigma \vdash B$ & $\Sigma \vdash \Diamond B$
(5)	t, t_2, f, F	$\Sigma \not\vdash \Box C$	t	$\Sigma \not\vdash \Box D$ & $\Sigma \vdash D$ & $\Sigma \vdash \Diamond D$	T, t	$\Sigma \vdash B$ & $\Sigma \vdash \Diamond B$
(6)	T, t_1, f_2, f_1	$\Sigma \vdash \Box C$	t_2	$\Sigma \not\vdash \Box D$ & $\Sigma \vdash D$ & $\Sigma \not\vdash \Diamond D$	t_2	$\Sigma \not\vdash \Box B$ & $\Sigma \vdash B$ & $\Sigma \not\vdash \Diamond B$
(7)	t, t_2, f, F	$\Sigma \not\vdash \Box C$	t_2	$\Sigma \not\vdash \Box D$ & $\Sigma \vdash D$ & $\Sigma \not\vdash \Diamond D$	T, t	$\Sigma \vdash B$ & $\Sigma \vdash \Diamond B$
(8)	$\mathcal{D}$	$\Sigma \vdash C$	f_2	$\Sigma \vdash \Box D$ & $\Sigma \not\vdash D$ & $\Sigma \vdash \Diamond D$	f_2	$\Sigma \vdash \Box B$ & $\Sigma \not\vdash B$ & $\Sigma \vdash \Diamond B$
(9)	$\mathcal{D}^c$	$\Sigma \not\vdash C$	f_2	$\Sigma \vdash \Box D$ & $\Sigma \not\vdash D$ & $\Sigma \vdash \Diamond D$	T	$\Sigma \vdash \Box B$ & $\Sigma \vdash B$ & $\Sigma \vdash \Diamond B$
(10)	T, t_1	$\Sigma \vdash \Box C$ & $\Sigma \vdash C$	f	$\Sigma \not\vdash \Box D$ & $\Sigma \not\vdash D$ & $\Sigma \vdash \Diamond D$	f	$\Sigma \not\vdash \Box B$ & $\Sigma \not\vdash B$ & $\Sigma \vdash \Diamond B$
(11)	t, t_2	$\Sigma \not\vdash \Box C$ & $\Sigma \vdash C$	f	$\Sigma \not\vdash \Box D$ & $\Sigma \not\vdash D$ & $\Sigma \vdash \Diamond D$	f_2, f	$\Sigma \not\vdash B$ & $\Sigma \vdash \Diamond B$
(12)	f_2, f_1	$\Sigma \vdash \Box C$ & $\Sigma \not\vdash C$	f	$\Sigma \not\vdash \Box D$ & $\Sigma \not\vdash D$ & $\Sigma \vdash \Diamond D$	t	$\Sigma \not\vdash \Box B$ & $\Sigma \vdash B$ & $\Sigma \vdash \Diamond B$
(13)	f, F	$\Sigma \not\vdash \Box C$ & $\Sigma \not\vdash C$	f	$\Sigma \not\vdash \Box D$ & $\Sigma \not\vdash D$ & $\Sigma \vdash \Diamond D$	T, t	$\Sigma \vdash B$ & $\Sigma \vdash \Diamond B$
(14)	T, t_1	$\Sigma \vdash \Box C$ & $\Sigma \vdash C$	f_1	$\Sigma \vdash \Box D$ & $\Sigma \not\vdash D$ & $\Sigma \not\vdash \Diamond D$	f_1	$\Sigma \vdash \Box B$ & $\Sigma \not\vdash B$ & $\Sigma \not\vdash \Diamond B$
(15)	t, t_2	$\Sigma \not\vdash \Box C$ & $\Sigma \vdash C$	f_1	$\Sigma \vdash \Box D$ & $\Sigma \not\vdash D$ & $\Sigma \not\vdash \Diamond D$	f_2	$\Sigma \vdash \Box B$ & $\Sigma \not\vdash B$ & $\Sigma \vdash \Diamond B$
(16)	f_2, f_1	$\Sigma \vdash \Box C$ & $\Sigma \not\vdash C$	f_1	$\Sigma \vdash \Box D$ & $\Sigma \not\vdash D$ & $\Sigma \not\vdash \Diamond D$	t_1	$\Sigma \vdash \Box B$ & $\Sigma \vdash B$ & $\Sigma \not\vdash \Diamond B$
(17)	f, F	$\Sigma \not\vdash \Box C$ & $\Sigma \not\vdash C$	f_1	$\Sigma \vdash \Box D$ & $\Sigma \not\vdash D$ & $\Sigma \not\vdash \Diamond D$	T	$\Sigma \vdash \Box B$ & $\Sigma \vdash B$ & $\Sigma \vdash \Diamond B$
(18)	T, t_1	$\Sigma \vdash \Box C$ & $\Sigma \vdash C$	F	$\Sigma \not\vdash \Box D$ & $\Sigma \not\vdash D$ & $\Sigma \not\vdash \Diamond D$	F	$\Sigma \not\vdash \Box B$ & $\Sigma \not\vdash B$ & $\Sigma \not\vdash \Diamond B$
(19)	t, t_2	$\Sigma \not\vdash \Box C$ & $\Sigma \vdash C$	F	$\Sigma \not\vdash \Box D$ & $\Sigma \not\vdash D$ & $\Sigma \not\vdash \Diamond D$	f_2, f	$\Sigma \not\vdash B$ & $\Sigma \vdash \Diamond B$
(20)	f_2, f_1	$\Sigma \vdash \Box C$ & $\Sigma \not\vdash C$	F	$\Sigma \not\vdash \Box D$ & $\Sigma \not\vdash D$ & $\Sigma \not\vdash \Diamond D$	t_2	$\Sigma \not\vdash \Box B$ & $\Sigma \vdash B$ & $\Sigma \not\vdash \Diamond B$
(21)	f, F	$\Sigma \not\vdash \Box C$ & $\Sigma \not\vdash C$	F	$\Sigma \not\vdash \Box D$ & $\Sigma \not\vdash D$ & $\Sigma \not\vdash \Diamond D$	T, t	$\Sigma \vdash B$ & $\Sigma \vdash \Diamond B$

By induction hypothesis, we have the conditions for C and D, and we can see that the conditions for B i.e. $C{\to}D$ are provable as follows:

For (1): (LK6), (Ax1), (LK4).   For (8): (LK6), (4), (LK4).   For (15): (LK6), (4), (LK3).
For (2): (LK6), (Ax1), (LK2).   For (9): (LK6), (5), (LK4).   For (16): (LK6), (5), (LK2).
For (3): (LK6), (Ax1), (LK3).   For (10): (LK), (4), (LK4).   For (17): (LK6), (5), (LK3).
For (4): (LK), (Ax1), (LK4).   For (11): (4), (LK4).   For (18): (LK), (4), (LK2).
For (5): (Ax1), (LK3).   For (12): (LK), (5), (LK4). For (19): (4), (LK3).
For (6): (LK), (Ax1), (LK2).   For (13): (5), (LK4).   For (20): (LK), (5), (LK2).
For (7): (Ax1), (LK3).   For (14): (LK6), (4), (LK2). For (21): (5), (LK3).

This completes the proof.

Proof for Lemma 71

Note first that v_0 is well-defined in view of (LT) and (5). Then the desired result can be proved by induction on the number n of connectives.

(Base): For atomic formulas, just note that Σ is consistent which implies that $\Sigma \vdash A$ or $\Sigma \vdash \neg A$.

(Induction step): We split the cases based on the connectives.

Case 1. If $B = \neg C$, then we have the following six cases.

cases	$v_\Sigma(C)$	condition for C	$v_\Sigma(B)$	condition for B i.e. $\neg C$
(i)	T	$\Sigma \vdash \Box C$	F	$\Sigma \vdash \Box \neg C$
(ii)	t_3	$\Sigma \not\vdash \Box C$ and $\Sigma \vdash \Diamond \Box C$	f_3	$\Sigma \not\vdash \Box \neg\neg C$ and $\Sigma \vdash \Diamond \Box \neg\neg C$ and $\Sigma \vdash \neg\neg C$
(iii)	t_4	$\Sigma \not\vdash \Box C$ and $\Sigma \not\vdash \Diamond \Box C$ and $\Sigma \vdash C$	f_4	$\Sigma \not\vdash \Box \neg\neg C$ and $\Sigma \not\vdash \Diamond \Box \neg\neg C$ and $\Sigma \vdash \neg\neg C$
(iv)	f_4	$\Sigma \not\vdash \Box \neg C$ and $\Sigma \not\vdash \Diamond \Box \neg C$ and $\Sigma \vdash \neg C$	t_4	$\Sigma \not\vdash \Box \neg C$ and $\Sigma \not\vdash \Diamond \Box \neg C$ and $\Sigma \vdash \neg C$
(v)	f_3	$\Sigma \not\vdash \Box \neg C$ and $\Sigma \vdash \Diamond \Box \neg C$ and $\Sigma \vdash \neg C$	t_3	$\Sigma \not\vdash \Box \neg C$ and $\Sigma \vdash \Diamond \Box \neg C$ and $\Sigma \vdash \neg C$
(vi)	F	$\Sigma \vdash \Box \neg C$	F	$\Sigma \vdash \Box \neg C$

By induction hypothesis, we have the conditions for C, and it is easy to see that the conditions for B i.e. $\neg C$ are provable. Indeed, (ii) and (iii) are provable in view of (10) and (11) respectively, and others immediate.

Case 2. If $B = \Box C$, then we have the following six cases.

cases	$v_\Sigma(C)$	condition for C	$v_\Sigma(B)$	condition for B i.e. $\Box C$
(i)	T	$\Sigma \vdash \Box C$	$\mathcal{D}$	$\Sigma \vdash \Box C$
(ii)	t_3	$\Sigma \not\vdash \Box C$ & $\Sigma \vdash \Diamond\Box C$ & $\Sigma \vdash C$	f_4, f_3	$\Sigma \not\vdash \Box\neg\Box C$ & $\Sigma \vdash \neg\Box C$
(iii)	t_4	$\Sigma \not\vdash \Box C$ & $\Sigma \not\vdash \Diamond\Box C$ & $\Sigma \vdash C$	F	$\Sigma \vdash \Box\neg\Box C$
(iv)	f_4	$\Sigma \not\vdash \Box\neg C$ & $\Sigma \not\vdash \Diamond\Box\neg C$ & $\Sigma \vdash \neg C$	F	$\Sigma \vdash \Box\neg\Box C$
(v)	f_3	$\Sigma \not\vdash \Box\neg C$ & $\Sigma \vdash \Diamond\Box\neg C$ & $\Sigma \vdash \neg C$	F	$\Sigma \vdash \Box\neg\Box C$
(vi)	F	$\Sigma \vdash \Box\neg C$	F	$\Sigma \vdash \Box\neg\Box C$

By induction hypothesis, we have the conditions for C, and we can see that the conditions for B i.e. $\Box C$ are provable. Indeed, (iii) is proved by (LM2), (iv) and (v) are immediate by (12), (vi) is provable by combining (LT) and (12), and others are obvious.

Case 3. If $B = \Diamond C$, then we have the following six cases.

cases	$v_\Sigma(C)$	condition for C	$v_\Sigma(B)$	condition for B i.e. $\Diamond C$
(i)	T	$\Sigma \vdash \Box C$	T	$\Sigma \vdash \Box\Diamond C$
(ii)	t_3	$\Sigma \not\vdash \Box C$ & $\Sigma \vdash \Diamond\Box C$ & $\Sigma \vdash C$	T	$\Sigma \vdash \Box\Diamond C$
(iii)	t_4	$\Sigma \not\vdash \Box C$ & $\Sigma \not\vdash \Diamond\Box C$ & $\Sigma \vdash C$	T	$\Sigma \vdash \Box\Diamond C$
(iv)	f_4	$\Sigma \not\vdash \Box\neg C$ & $\Sigma \not\vdash \Diamond\Box\neg C$ & $\Sigma \vdash \neg C$	T	$\Sigma \vdash \Box\Diamond C$
(v)	f_3	$\Sigma \not\vdash \Box\neg C$ & $\Sigma \vdash \Diamond\Box\neg C$ & $\Sigma \vdash \neg C$	t_3, t_4	$\Sigma \not\vdash \Box\Diamond C$ & $\Sigma \vdash \Diamond C$
(vi)	F	$\Sigma \vdash \Box\neg C$	$\mathcal{F}$	$\Sigma \vdash \neg\Diamond C$

By induction hypothesis, we have the conditions for C, and we can see that the conditions for B i.e. $\Diamond C$ are provable. Indeed, (i) is provable by combining (LT) and (LB), (ii) and (iii) are immediate by (LB), (iv) and (v) are provable by (B4) and (B3) respectively, and (vi) is provable by (LM1).

Case 4. If $B = C \to D$, then we have the following 11 cases.

cases	$v_\Sigma(C)$	condition for C	$v_\Sigma(D)$	condition for D	$v_\Sigma(B)$	condition for B i.e. $C \to D$
(i)	F	$\Sigma \vdash \Box\neg C$	any	—	T	$\Sigma \vdash \Box(C \to D)$
(ii)	any	—	T	$\Sigma \vdash \Box D$	T	$\Sigma \vdash \Box(C \to D)$
(iii)	T	$\Sigma \vdash \Box C$	t_3, t_4	$\Sigma \not\vdash \Box D$ & $\Sigma \vdash D$	t_3, t_4	$\Sigma \not\vdash \Box(C \to D)$ & $\Sigma \vdash (C \to D)$
(iv)	T	$\Sigma \vdash \Box C$	f_4, f_3	$\Sigma \not\vdash \Box\neg D$ & $\Sigma \vdash \neg D$	f_4, f_3	$\Sigma \not\vdash \Box\neg(C \to D)$ & $\Sigma \vdash \neg(C \to D)$
(v)	T	$\Sigma \vdash \Box C$	F	$\Sigma \vdash \Box\neg D$	F	$\Sigma \vdash \Box\neg(C \to D)$
(vi)	t_3, t_4	$\Sigma \not\vdash \Box C$ & $\Sigma \vdash C$	t_3, t_4	$\Sigma \not\vdash \Box D$ & $\Sigma \vdash D$	$\mathcal{D}$	$\Sigma \vdash C \to D$
(vii)	t_3, t_4	$\Sigma \not\vdash \Box C$ & $\Sigma \vdash C$	f_4, f_3	$\Sigma \not\vdash \Box\neg D$ & $\Sigma \vdash \neg D$	f_4, f_3	$\Sigma \not\vdash \Box\neg(C \to D)$ & $\Sigma \vdash \neg(C \to D)$
(viii)	t_3, t_4	$\Sigma \not\vdash \Box C$ & $\Sigma \vdash C$	F	$\Sigma \vdash \Box\neg D$	f_4, f_3	$\Sigma \not\vdash \Box\neg(C \to D)$ & $\Sigma \vdash \neg(C \to D)$
(ix)	f_4, f_3	$\Sigma \not\vdash \Box\neg C$ & $\Sigma \vdash \neg C$	t_3, t_4	$\Sigma \not\vdash \Box D$ & $\Sigma \vdash D$	$\mathcal{D}$	$\Sigma \vdash C \to D$
(x)	f_4, f_3	$\Sigma \not\vdash \Box\neg C$ & $\Sigma \vdash \neg C$	f_4, f_3	$\Sigma \not\vdash \Box\neg D$ & $\Sigma \vdash \neg D$	$\mathcal{D}$	$\Sigma \vdash C \to D$
(xi)	f_4, f_3	$\Sigma \not\vdash \Box\neg C$ & $\Sigma \vdash \neg C$	F	$\Sigma \vdash \Box\neg D$	t_3, t_4	$\Sigma \not\vdash \Box(C \to D)$ & $\Sigma \vdash (C \to D)$

By induction hypothesis, we have the conditions for C and D, and we can see that the conditions for B i.e. $C \to D$ are provable as follows:

- For (i) and (ii), use (LK5) and (LK6) respectively.
- For (iii), $\Sigma \vdash C \to D$ follows immediately by $\Sigma \vdash D$ and (Ax1). For the other half, assume $\Sigma \vdash \Box C$ and $\Sigma \not\vdash \Box D$. Then the latter is equivalent to $\Sigma \vdash \neg\Box D$, and so in view of (1), we obtain $\Sigma \vdash \neg\Box(C \to D)$ i.e. $\Sigma \not\vdash \Box(C \to D)$, as desired.
- For (iv), $\Sigma \vdash \neg(C \to D)$ follows in view of (4). For the other half, by (LK4).

- For (v), assume $\Sigma \vdash \Box C$ and $\Sigma \nvdash \Box\neg(C\to D)$. Then by the latter we have $\Sigma \vdash \Diamond(C\to D)$. Therefore, in view of (LK2), we obtain $\Sigma \vdash \Diamond D$, i.e. $\Sigma \nvdash \Box\neg D$, as desired.
- For (vi) and (ix), just use (Ax1).
- For (vii) and (viii), $\Sigma \vdash \neg(C\to D)$ follows in view of (4). For the other half, by (LK3).
- For (x), just use (5).
- For (xi), $\Sigma \vdash C\to1 D$ follows in view of (Ax1). For the other half, by (LK1).

This reuses the proof in the case of the modal logic **T**, by splitting **t** and **f** into $\mathbf{t_3}, \mathbf{t_4}$ and $\mathbf{f_4}, \mathbf{f_3}$ respectively. This completes the proof.

Three Grades of Agnostic Involvement

GILLMAN PAYETTE
University of British Columbia

Abstract

In this paper I explore agnosticism from an epistemic logic perspective. The view of agnosticism I take is that one is agnostic about a proposition when one believes that proposition to be consistent with what one knows. I look at how various assumptions about epistemic and doxastic logic interact with ways that one can formulate agnosticism given the basic intuitive reading. I focus on agnosticism about certain philosophical and theological positions that can be formulated in terms of all true propositions being such and such. For example, God being the cause of all truths. I introduce modal operators to formulate such claims and then formulate what it is to be agnostic about such claims. I show that three formulations of agnosticism, one which is the basic formulation and two which formulate agnosticism as a position of rational fairness, result in problematic claims on just about all common formulations of epistemic and doxastic logic. Essentially, what is shown is that one cannot be agnostic about determinism, for example, while entertaining the hypothesis that particular propositions are contingently true, e.g., I didn't have to wear a black t-shirt today.

1 Introduction

In this paper I will consider agnosticism, in a general sense, from the standpoint of epistemic logic. I will argue that there is tension when we try to put agnosticism into wide reflective equilibrium with commitments of ideal rationality as represented by (classical) epistemic logic. The result, however, is not that one is rationally committed to belief in philosophical or theological claims, but rather that the agnostic position about such claims is (conceptually) problematic *under those assumptions*.

In what follows I will first have a brief discussion about what I mean by 'agnosticism', then, in section 2, discuss the assumptions of epistemic and doxastic logic that

---

The author would like to thank John Woods for reading earlier drafts of this paper, and Travis Dumsday for some discussions of the nature of God, as well as the anonymous referees for their comments. The author is also thankful to the Social Sciences and Humanities Research Council of Canada for funding this research through a Banting Postdoctoral Fellowship.

the rest of the essay will employ. In section 3, I will move to a formal representation of the conceptions of agnosticism, and then show how those representations lead to—perhaps—unwanted results in section 4. In the end, what we can see is that agnosticism is incompatible with belief or disbelief in certain philosophical/theological views under the assumptions of logical omniscience and positive introspection. My results can be seen more as a commentary on the feasibility of representing agnosticism in epistemic/doxastic logic than a project in logical philosophy of religion/metaphysics. I will discuss this in more detail in section 6.

Here I will return to Huxley's Greek root of the term 'agnosticism' in the sense of 'not known'. A general explanation of agnosticism is, as Pojman puts it, the view that "metaphysical ideas cannot be proved or disproved" [12, p. 15]. But contemporary analytic metaphysics would disagree with such a sweeping skepticism; we have come a long way since the logical positivism of the 1930s. If we are agnostic about some metaphysical notion these days, it seems that we are willing to agree that, at least, propositions about those ideas make sense. We are just unwilling to say that we *know* whether they are true. The sweeping conception of agnosticism is now referred to as 'strong agnosticism': one *cannot* know something (say φ).[1] The sense of agnosticism that I consider in this paper would be so-called 'weak, local agnosticism'. It is weak in that it doesn't deny the possibility of knowledge of φ, and it is local since it only concerns knowledge of that φ, not a whole field of inquiry—cf. Poidevin [11, Ch. 1].

On this view of agnosticism I am not claiming that agnosticism about φ is a position in a debate about whether φ, it is more a position about one's epistemic situation relative to φ. I claim that being agnostic in this weak local sense is an attitude, but it is not an attitude toward φ.[2] Being agnostic is something an agent *does*. However, an agnostic needn't be perpetually upholding that attitude. An agnostic can be disposed toward agnosticism, and thereby be agnostic. I will explain this latter position on agnosticism in section 3 in relation to the formulation of agnosticism I call 3A.

---

[1]Rosenkranz [13] refers to two kinds of strong agnosticism as levels of '*True Agnosticism*'(p. 99). The first level means that we will not be in a position to know a proposition (φ) which we do not know whether it is true, given our current understanding of what it is to know something. Rosenkranz's idea is that sometimes in order to advance knowledge there must be a radical shift in how things are known. First level agnosticism doesn't deny that knowledge on the subject will ever be had; a radical change in how we know things may be needed, cf. Kuhn [10]. The second level true agnostic denies that we can ever be in a position to know (given our current way of knowing) that we could ever know—radical change or not—whether we could know whether φ is true. I am not interpreting agnosticism in either of these ways.

[2]In claiming this I am setting my position apart from that of Rosenkranz [13] who is concerned to show that agnosticism can be a "third position" in debates about whether φ.

My final claim is that agnosticism regarding φ should be consistent with *beliefs* in propositions which would decide whether φ. I think that kind of consistency should be a requirement of theoretical inquiry. We consider propositions, decide whether they are plausible, and believe them. But in believing a proposition, particularly in theoretical inquiry, the responsible theorist recognizes a difference between rational belief and knowledge. There is yet another position that one might take toward φ: subjective certainty. When one is subjectively certain, one may find evidence for φ very compelling, and on that basis claim knowledge of φ. But one may be wrong about that. That is different from belief, and also different from recognizing that one is rational in a belief. And both are different from actually *knowing* φ. The logic I will develop below attempts at keeping these three stances separate.

Agnosticism on this view can play another role in inquiry. If one is to take a philosophical, scientific, theological... position seriously, they must not be subjectively certain of propositions which are contrary to that position. This is what I will call *rational fairness*. I think that rational fairness is a form of agnosticism. However, I also think rational fairness could be logically independent from the weak local agnosticism discussed above. These variations on agnosticism will lead to three grades of agnosticism which I will present in section 3.

2 Some Formalities

The background logic that I am assuming is classical propositional logic augmented with propositional quantifiers. Thus, there are the usual boolean connectives $\&, \vee, \sim, \leftrightarrow, \to$, and the propositional quantifiers: $\forall, \exists$. The formulas can then be written as $\varphi(p_1, \ldots, p_n)$ where the p_i are propositional variables or atoms in the formula. The expression $\varphi(p_1, \ldots, \theta/p_i, \ldots p_n)$ then is used to express the formula φ with θ uniformly substituted for p_i. Here I am using lowercase Greek letters for formula meta-variables; the lowercase Latin letters are propositional variables. The rules/axioms for the quantifiers are the usual ones from classical first-order logic:

$$\frac{\varphi \to \psi(p_1, \ldots, p_i, \ldots p_n)}{\varphi \to (\forall p_i)\psi(p_1, \ldots, p_i, \ldots p_n)} \, [\forall I] \quad \text{provided that } p_i \text{ doesn't occur free in } \varphi,$$

$$(\forall p_i)\varphi(p_1, \ldots, p_i, \ldots p_n) \to \varphi(p_1, \ldots, \theta/p_i, \ldots p_n) \, [\forall E]$$

$$\psi(p_1, \ldots, \theta/p_i, \ldots p_n) \to (\exists p_i)\psi(p_1, \ldots, p_i, \ldots p_n) \, [\exists I],$$

$$\frac{\varphi(p_1, \ldots, p_i, \ldots p_n) \to \psi}{(\exists p_i)\varphi(p_1, \ldots, p_i, \ldots p_n)\varphi \to \psi} \, [\exists E] \quad \text{provided that } p_i \text{ doesn't occur free in } \psi$$

Uniform Substitution [US]

Logics like these can be found in Bull [3] and Antonelli and Thomason [1]. Uniform substitution must be put under an important restriction. We do not want free occurrences of a propositional variable in a formula φ to become bound when substituted into another formula ψ. Thus, we will make sure that doesn't happen. It is important to note that complex formulas can only be quantified away in the case of ∀E and ∃I. In the case of the other two quantifier rules, the formulas must be propositional variables.

My discussion of agnosticism hinges on the concepts of knowledge and belief. Thus I will add the operators for those concepts to the language: K and B. Usually these are agent relative, but for the most part I will leave reference to the agent implicit. In the following subsections I discuss the logical properties each of these operators might have. In section 2.3 I summarize the properties of the operators which I discuss in the next three subsections.[3]

2.1 Knowledge

Epistemic logic is a tricky subject, and has been since its beginnings in Hintikka [8]. A common modal system for knowledge is S5. This is an assumption made by game theorists and theoretical computer scientists, cf. Aumann and Brandenburger [2]. According to just about any system of epistemic logic, our knowledge is closed under logical consequence, and even known implications. There have been reams of paper used in presenting arguments discussing the feasibility of the principles for epistemic logic. In this paper I will offer a specific, not novel, reading which may support the principles I propose. I am not terribly concerned with providing indefeasible support for the principles since they are commonly assumed and my goal is really to study the relationship between those principles (which formulate conceptions of ideal rationality) and the notions of agnosticism. However, they should contain some plausibility otherwise the arguments offered wouldn't be of much interest.

The operator that I will use in most of my formulations of agnosticism is the dual of the knowledge operator: $\langle K \rangle$ (which is defined as $\langle K \rangle =_{df} \sim K \sim$). It is often read as 'for all I know ... is the case'. But the way the operator should be interpreted in epistemic logic is literally: it is not the case that I know that it is not the case that .... That reading, when applied to $\langle K \rangle \varphi$, says 'I don't know $\neg \varphi$'. But that says nothing about my epistemic relationship to logical consequences of φ.

Some suggest that $\langle K \rangle$ should be read as: it is consistent with what I know that φ. To say that φ is consistent with what I know simply means that $\neg \varphi$ is not a logical consequence of what I know. But that interpretation gives a different gloss

---

[3]Also note that I am using the convention of interpreting the scope of modal operators as being narrow in the absence of parentheses.

to $K\varphi$; the latter should really be read as: it is a logical consequence of what I know that φ. It is this 'logical consequence' reading that I rely on since it describes the situation I am interested in.

Particularly, I am interested in epistemic and doxastic logic as models of rational investigation. They are descriptions of the logical connections between what one might call the 'conventional' beliefs and the logical commitments of those beliefs. Thus, I don't want to say $K\varphi$ is to mean that the agent actually knows the proposition in the more conventional sense where one is aware of φ. But rather that what the agent does know commits one to φ. Having said that, I will have to be careful not to read conventional renderings of natural language sentences into the formal language which are not supported by this less intuitive, albeit very popular, reading.

Given my (non-novel) reading of the K operator, the following are rules of the system:

$$\frac{\varphi \to \psi}{K\varphi \to K\psi} \; [RMK],$$

and

$$\frac{\varphi \leftrightarrow \psi}{\langle K \rangle \varphi \leftrightarrow \langle K \rangle \psi} \; [REK].$$

REK follows from the monotonicity condition RMK that I have suggested is permissible given my reading of K, but I mostly use REK in what follows. Another rule which follows from my reading of K is that all logical truths are known—which includes the theorems of the system of course. The reason for this is that logical truths are logical consequences of every proposition; thus, they will all be logical consequences of what I know, if I know anything at all. What this amounts to at the very least is that I will assume that one knows the basic logical truth: $\top$. Therefore,

$$K\top \; [NK].$$

It follows from NK using REK and the definition of $\langle K \rangle$ that:

$$\bot \leftrightarrow \langle K \rangle \bot \; [N\langle K \rangle]$$

which is what will feature prominently in the proofs to follow.

If $\varphi \to \psi$ is "known" and so is φ, then ψ is a logical consequence as well. Thus, the axiom:

$$K(\varphi \to \psi) \to (K\varphi \to K\psi) \; [RegK]$$

is valid. We also need some rules that allow the quantifiers and operators to interact.

I suggest the following:

$$(\exists p)K(\varphi) \to K((\exists p)\varphi) \ [\exists K]$$

and

$$K(\forall p)\varphi \to (\forall p)K\varphi \ [\forall K].$$

I think these rules fall out of my interpretation of K since if I know that φ, then it is a logical consequence of φ that $(\exists p)\varphi$. That is simply an application of standard logical rules for quantifiers. In the universal case, if I know that φ holds for all p, then it is a logical consequence of $(\forall p)\varphi$, that φ holds for any particular instance of p. Any instance is just a logical consequence of my knowledge, so it holds for all p. Finally, I will assume that knowledge is also factive; we cannot know false things. Thus,

$$K\varphi \to \varphi \ [TK].$$

There are a number of other assumptions that one could make about the knowledge operator, but they are often disputed. I will discuss two of them here: the so-called 'positive introspection' or KK principle, $K\varphi \to KK\varphi$, and its negative cousin. Positive introspection or PI says that if we know, we know that we know. It is a controversial principle. Hintikka [8] gives an interesting argument for it, but I want to offer another justification given the 'logical consequence' reading. This is perhaps unnecessary given the wide acceptance of the principle on the 'available information' reading of K which is more or less the same as my 'logical consequence' reading. I think that the PI principle can be seen as the expression of the deduction theorem. Consider what $KK\varphi$ says. It says that $K\varphi$ follows from what one knows. Expanding out further: it follows from what one knows that φ follows from what one knows. If φ does follow logically from one's knowledge (call that collection of propositions Γ_K), then one's knowledge implies that proposition, i.e., $\Gamma_K \vdash \varphi$. By the deduction theorem, then, $\vdash \bigwedge \Gamma_K \to \varphi$. Given the assumption that one knows $\top$, and that $\bigwedge \Gamma_K \to \varphi$ is equivalent to $\top$, $K(\bigwedge \Gamma_K \to \varphi)$ is also a truth of epistemic logic. But that is to say that it follows logically from what one knows that φ follows from what one knows, which is what $KK\varphi$ says. It seems to follow from the assumption of logical omniscience. Thus, I will make PI part of my basic system.[4]

---

[4] As an anonymous reviewer reminded me, adding a theory like Robinson arithmetic to a modal logic like the one I have defined is problematic because of Löb's theorem, cf. Smorynski [17]. Such logics can translate the sentences of the language via Gödel numberings, but then sentences like $K\varphi \to \varphi$ generate inconsistencies because of Gödel's incompleteness theorems. The same effects can be reproduced in modal languages if one adds fixed point axioms for each formula: $\varphi(\delta_\varphi) \longleftrightarrow \delta_\varphi$ for each formula φ. This mimics the result of the diagonalization theorem provable in Robinson arithmetic. What one can then show, a result due to Montague, is that modal logics that have

The principle that is usually called 'negative introspection' (NI) assumes that if one doesn't know something, then one *knows* that one lacks that knowledge. This is usually formulated as:
$$\sim K\varphi \to K \sim K\varphi.$$
This axiom is equivalent to the 5 axiom $\langle K \rangle \varphi \to K \langle K \rangle \varphi$, and it is also equivalent to $\langle K \rangle K\varphi \to K\varphi$. I don't think that it can be given the same kind of rationale on my reading of K as PI. Since $K\varphi$ means that φ is a logical consequence of one's knowledge, $\sim K\varphi$ means that φ isn't a logical consequence of one's knowledge. In the case of positive introspection, one can reflect $\Gamma_K \vdash \varphi$ into the object language via the deduction theorem. In the case of $\Gamma_K \nvdash \varphi$ the same reflection isn't possible. While the truth of $\sim (\wedge\Gamma_K \to \varphi)$ would mean that $\Gamma_K \nvdash \varphi$, and $\sim (\wedge\Gamma_K \to \varphi)$ should be a priori true, the *logical* truth of $\sim (\wedge\Gamma_K \to \varphi)$ is much stronger than what $\Gamma_K \nvdash \varphi$ means. This principle was rejected by Hintikka and others, but it can have interesting consequences, so although I will not make it part of my basic system, I want to explore those consequences.

2.2 Beliefs

Traditionally, i.e., from Hintikka [8], the logic of belief is almost as strong as that of knowledge. Factivity ($B\varphi \to \varphi$) is all that is denied to it. Again, I will take the 'is a logical consequence of what I believe' or ideally rational agent reading for the operator B. That means $\langle B \rangle$ is to be read as 'is logically consistent with what I believe' as well. This reading allows me to use similar rules to those for knowledge to describe the logical properties of belief. Particularly the following:

$\bot \longleftrightarrow \langle B \rangle \bot$ [N$\langle B \rangle$]

$\dfrac{\varphi \to \psi}{B\varphi \to B\psi}$ [RMB]

$\dfrac{\varphi \longleftrightarrow \psi}{\langle B \rangle \varphi \longleftrightarrow \langle B \rangle \psi}$ [REB]

$B(\varphi \to \psi) \to (B\varphi \to B\psi)$ [RegB]

$(\exists p)B(\varphi) \to B((\exists p)\varphi)$ [∃B]

---

necessitation and T are inconsistent, cf. Stern and Fischer [19]. These results indicate that any modal logic which contains such formulas and rules is problematic. If I were really trying to defend PI and translate formulas into terms, I would have to use Gödel numberings, but I am really just trying provide some rationale for PI, rather than mounting a defense of it.

$$B(\forall p)\varphi \to (\forall p)B\varphi \ [\forall B]$$

and all for the same reasons as in the case of knowledge. The special assumptions that I will make for belief are as follows. First, we need to connect belief and knowledge. To know something, one must believe it. Thus, if something follows from one's knowledge, then it must also follow from one's beliefs. Therefore, we can assume that knowledge implies belief, i.e.,

$$K\varphi \to B\varphi \ [KB].$$

It follows immediately via the definitions of $\langle K \rangle$ and $\langle B \rangle$, that $\langle B \rangle \varphi \to \langle K \rangle \varphi$.

The tricky assumption about belief, which comes for free in the case of knowledge, is that one's beliefs are consistent:

$$B\varphi \to \langle B \rangle \varphi \ [DB].$$

This schema is called DB because it is a version of the D axiom for doxastic logic or the logic of belief. It is acceptable in the circumstances that I think are relevant to my discussion. These are, after all, idealizing assumptions.

There are other interaction axioms suggested by Stalnaker [18] that should be considered as well. For example, if one believes something, then one knows that one does: $B\varphi \to KB\varphi$. This is acceptable, if we can assume that we have full knowledge of our positive doxastic and epistemic states. Of course it can be given a similar justification as PI above. It follows, then, that belief implies belief of belief: $B\varphi \to BB\varphi$. I think this BB condition is acceptable for the same reason as positive introspection as well.

There is a corresponding negative introspection condition which combines the two operators: $\sim B\varphi \to K \sim B\varphi$ [BKB]. It says that one has full epistemic access to one's doxastic state, but we can still be ignorant of things we do not know. Allowing that kind of ignorance is important because even the logically perfect might still have false beliefs. This axiom doesn't present the same problems as NI, but it is not sensible on my reading of the operators given what I have said about NI above.

Despite NI being questionable on the reading I have offered, the major problem I see with NI is that mere belief becomes too powerful; we can come to know things by lying to ourselves. By assuming NI, believing one knows something implies that one actually knows that thing, i.e., $BK\varphi \to K\varphi$ cf. Stalnaker [18, p. 179]. That follows since $BK\varphi$ implies $\langle B \rangle K\varphi$ by DB, and the contrapositive of KB allows $\langle B \rangle K\varphi$ to imply $\langle K \rangle K\varphi$. NI finally allows us to conclude $K\varphi$ from $BK\varphi$. Indeed, being strongly convinced of something would make it true, and that is beyond the pale. Of course, it is unproblematic to assume that if one believes that one knows, then one believes: $BK\varphi \to B\varphi$.

Finally, there is what Stalnaker calls the axiom of strong belief: $B\varphi \to BK\varphi$. This axiom is completely at odds with the agnostic position I am considering; Stalnaker sees this as an expression of subjective certainty. It doesn't permit the believer any humility, epistemically speaking. For the agnostic, they may believe something, but recognize that they do not *know* all of the propositions that they believe. On some readings of belief this may be acceptable, but not on mine. I should be able to say I believe something, and so take it to be true, but recognize that I shouldn't claim it as knowledge since I recognize it as revisable. The reason is that the beliefs may not be based on very much: $B\varphi$ is a mere belief in φ or is the logical consequence of the things I know and merely believe. The important restriction is that the agent recognizes that its beliefs might be wrong. On a technical note, it will be inconsistent for an agent to be subjectively certain about φ, but also believe that $\sim \varphi$ is consistent with what it knows. Thus, $BK\varphi \to \sim B\langle K\rangle \sim \varphi$ is a consequence of the system.

An important point to take from these discussions is that axioms which embed operators don't need to be interpreted in terms of an agent being aware of the propositions that follow the operators; the agent doesn't need to be aware that they believe φ when $BB\varphi$ is true. What $BB\varphi$ expresses is that the agent's attitude of belief can be extended, logically, to $B\varphi$, and that can happen when the agent's beliefs can be extended logically to φ. Even if you are inclined to reject all of my justifications for the axioms, I could construe my results as a test of common assumptions made in doxastic and epistemic logic when applied to this particular problem. Whichever of those attitudes you take toward the axioms leaves the results untouched. Before I return to agnosticism I will summarize the logical system for knowledge and belief.

2.3 Summary

I will call my basic system for knowledge and belief L_{KB}, and I present a more compact description of the axioms and rules here than what was elaborated above. The readings of the K and B operators are: 'it is a logical consequence of what I know that ...' and 'it is a logical consequence of what I believe that...'. In the next section I will introduce a new operator ■ to formalize the agnostic positions I want to discuss. There I will offer two logical principles for it, RM■ and T■, which will complete the system. Given these readings, I believe the following axioms and rules are defensible for the system:

- Classical propositional logic [PL], and propositional quantifiers:

$$\frac{\varphi \to \psi(p_1,\ldots,p_i,\ldots p_n)}{\varphi \to (\forall p_i)\psi(p_1,\ldots,p_i,\ldots p_n)} \ [\forall I] \quad \text{provided that } p_i \text{ doesn't occur free in } \varphi,$$

$$(\forall p_i)\varphi(p_1,\ldots,p_i,\ldots p_n) \to \varphi(p_1,\ldots,\theta/p_i,\ldots p_n) \ [\forall E]$$

$$\psi(p_1,\ldots,\theta/p_i,\ldots p_n) \to (\exists p_i)\psi(p_1,\ldots,p_i,\ldots p_n) \ [\exists I],$$

$$\frac{\varphi(p_1,\ldots,p_i,\ldots p_n) \to \psi}{(\exists p_i)\varphi(p_1,\ldots,p_i,\ldots p_n)\varphi \to \psi} \ [\exists E] \quad \text{provided that } p_i \text{ doesn't occur free in } \psi$$

Uniform Substitution [US]

- Pure Knowledge obeys the rules and axioms RMK, TK, PI, NK, RegK, ∀K and ∃K (the logic is roughly S4 with quantifiers):

$$\frac{\varphi \to \psi}{K\varphi \to K\psi} \ [RMK]$$

$K\varphi \to \varphi$ [TK]

$K\top$ [NK]

$K\varphi \to KK\varphi$ [PI]

$K(\varphi \to \psi) \to (K\varphi \to K\psi)$ [RegK]

$(\exists p)K(\varphi) \to K((\exists p)\varphi)$ [∃K]

$K(\forall p)\varphi \to (\forall p)K\varphi$ [∀K]

- Pure Belief obeys the rules and axioms NecB, RegB, DB, BB, ∀B and ∃B (the logic is roughly KD4 with quantifiers):

$B(\varphi \to \psi) \to (B\varphi \to B\psi)$ [RegB]

$B\varphi \to \langle B \rangle \varphi$ [DB]

$B\varphi \to BB\varphi$ [BB]

$\dfrac{\varphi}{B\varphi}$ [NecB]

$(\exists p)B(\varphi) \to B((\exists p)\varphi)$ [∃B]

$B(\forall p)\varphi \to (\forall p)B\varphi$ [∀B]

- Interaction Axioms:

$K\varphi \to B\varphi$ [KB]

$B\varphi \to KB\varphi$ [BKB]

That is the basic system L_{KB}. There are other axioms which I think are unacceptable, but have been accepted in some cases. These axioms are interesting, so should be used to test the limits of my results below:

$\sim B\varphi \to K \sim B\varphi$ [NIB]

$\sim K\varphi \to K \sim K\varphi$ [NI]

$B\varphi \to BK\varphi$ [SB]

3 Three Grades of Agnosticism

Before coming to a formal account of agnosticism, I will recapitulate what I mean by that term. Recall that agnosticism, as I am intending it, and as it seems Huxley intended it, is about knowledge. Thus being agnostic has to do with an epistemic state of an agent toward a proposition which one is agnostic about. The propositions I will be interested in for the current paper all have a certain form which I will define after discussing this general notion of agnosticism.

Agnosticism about a proposition, generally, is being in a position where one doesn't know whether that proposition is true. One knows whether p is true iff one knows that p or one knows that not-p, i.e., $Kp \vee K \sim p$. Not knowing whether p, then, is formalized as: $\sim (K \sim p \vee Kp)$, or, equivalently, $\sim K \sim p \ \& \ \sim Kp$. This position can be given another gloss given the formalism of this paper. To say that one is agnostic about p is to say that it is consistent with what one knows that p, but also that it is consistent with what one knows that not-p ($\sim p$). Naturally, then, one is tempted to formalize agnosticism about p as $\langle K \rangle p \ \& \ \langle K \rangle \sim p$, but I want to reflect a moment on the agnostic's position.

In saying that one is agnostic, one is saying that they do not *know* which proposition to hold: p or $\sim p$, both seem tenable. Agnosticism about a proposition is an attitude towards that proposition. The sentence $\langle K \rangle p \ \& \ \langle K \rangle \sim p$ expresses the *absence* of an attitude toward p, and $\sim p$. Hence, a formalization of an agnostic's claims should also express some attitude. What $\langle K \rangle p \ \& \ \langle K \rangle \sim p$ does provide for is the tenability of both options; the agent doesn't know anything which will rule one of them out. Indeed, $\langle K \rangle p \ \& \ \langle K \rangle \sim p$ means there is in fact nothing in what the agent knows that can rule one of the propositions out; it is more than a mere seeming.

The way I suggest formalizing agnosticism is to represent the agent as having an attitude towards the fact that both propositions are tenable. Thus, there are two candidates for how to formalize agnosticism: $K(\langle K\rangle p \;\&\; \langle K\rangle \sim p)$ and $B(\langle K\rangle p \;\&\; \langle K\rangle \sim p)$. Given L_{KB}, the first implies the second. Since it is more intuitive to capture 'seeming' via belief than knowledge, I suggest using the second option. Some may still object that agnosticism is about belief, so say that it should be captured by $\langle B\rangle p \;\&\; \langle B\rangle \sim p$. But I have three things to say about that. First, by the contrapositive of KB, $\langle B\rangle p \;\&\; \langle B\rangle \sim p$ implies $\langle K\rangle p \;\&\; \langle K\rangle \sim p$. So I am just taking the weakest form of agnosticism that makes sense as my starting point. Second, as I suggested above, proper philosophical methodology must allow us to take up beliefs while still remaining open-minded. We should be able to make hypotheses, and then reject them at a later time given new evidence or argument. If agnosticism is not believing whether φ, then believing φ would make one non-agnostic. Third, the knowledge reading is in line with the literature on this subject, e.g., Poidevin [11] and Rosenkranz [13].

Thus, an agent is agnostic about p iff $B(\langle K\rangle p \;\&\; \langle K\rangle \sim p)$ is true. Agnosticism is a second order attitude, it is a belief about our epistemic state. On a technical point $B(\langle K\rangle p \;\&\; \langle K\rangle \sim p)$ is equivalent to $B\langle K\rangle p \;\&\; B\langle K\rangle \sim p$ since B is a normal modal operator. Via DB and the contrapositives of KB and PI it will follow that $B(\langle K\rangle p \;\&\; \langle K\rangle \sim p)$ implies $\langle K\rangle p \;\&\; \langle K\rangle \sim p$. So even in the case of believing a proposition and its negation are consistent with what one knows, there is no mere seeming, at least in L_{KB}.

I am interested in agnosticism about certain kinds of propositions all of which can be expressed in terms of certain modal formulas. For this purpose I will introduce a third modal operator ∎. The kinds of principles I will assume are the following two:

$$\frac{\varphi \to \psi}{\blacksquare\varphi \to \blacksquare\psi} \; [RM\blacksquare]$$

and says that ∎ is closed under logical consequence. The following rule, which allows the substitution of logical equivalents into the ∎ operator, follows from RM∎:

$$\frac{\varphi \leftrightarrow \psi}{\blacksquare\varphi \leftrightarrow \blacksquare\psi} \; [RE\blacksquare],$$

as does the schema:

$$\blacksquare(\psi \;\&\; \varphi) \to (\blacksquare\psi \;\&\; \blacksquare\varphi) \; [Dist].$$

Next we have that ■ is a factive operator.

$$■\varphi \to \varphi \quad [T■].$$

I choose T■ for its name to indicate that it is the T axiom for the ■ operator. Note that neither of these principles imply that ■φ is logically true for any φ. The logical system L_{KB} extended with these two rules will be denoted $L_{KB}^{■}$. Next, I will take up the various interpretations that one might give for ■, but postpone discussion of whether these interpretations satisfy these principles in section 6.

Turning to the propositions I am interested in, each has the form $(\forall p)(p \to ■p)$. This is to say that for any proposition, if it is the case, then it is the case that ■ holds of that proposition as well. A number of philosophical positions can be expressed this way. Of the three that I will consider two involve the powers of God and one involves necessity. The powers of God are God's omniscience, and what I will call omnipotence although I am giving it a rather non-standard reading. The sense that I am giving to 'omnipotence' is that God is the truthmaker of all propositions which obtain. Those who think we have freewill would probably be skeptical of such a position. If ■ is necessity, then $(\forall p)(p \to ■p)$ is an expression of a kind of determinism: all true propositions are necessarily so. Other interpretations are certainly possible, but for the moment I will focus mostly on formal results which arise from expressing agnosticism about these positions. I will make reference to these interpretations to understand what the grades of agnosticism might mean.

Agnosticism about $(\forall p)(p \to ■p)$ would be rendered in what I call the first grade of agnostic involvement, which is the sense I discussed above, as follows:

Definition 1 (First Grade of Agnostic Involvement). An agent a is *agnostic* (first grade) iff

$$(A) \; B_a \langle K_a \rangle (\forall p)(p \to ■p) \text{ and } (\Omega) \; B_a \langle K_a \rangle (\exists p)(p \; \& \sim ■p)$$

Here the subscript 'a' indicates that a has the relevant attitude. In the rest of the paper it is assumed that a is constant and implicit.

This kind of agnosticism commits one to very little, but it commits one to *something*. It is a recognition that one's knowledge is limited in a certain way. It should leave room for one to *believe* or disbelieve $(\forall p)(p \to ■p)$. If I claimed to *know* one of $(\forall p)(p \to ■p)$ or $(\exists p)(p \; \& \sim ■p)$, then I don't think one could say I was agnostic, but it does seem consistent to say that while I am agnostic about $(\forall p)(p \to ■p)$, I *believe* that—for example—God's power satisfies $(\forall p)(p \to ■p)$.

If ■ is necessity, metaphysical or otherwise, then $(\forall p)(p \to ■p)$ says all propositions which are the case, are so necessarily. Agnosticism about that proposition means that it is an epistemic possibility that there are contingently true propositions, but that it is also an epistemic possibility that all true propositions are necessary. Again, it is an epistemic possibility that some evidence—in a wide sense—might be found that would sway the agnostic one way or the other, but the agnostic doesn't have that evidence.

The agnostic in the next version has an attitude of rational fairness toward $\blacksquare p$ and $p\ \&\ \sim \blacksquare p$. Like Jason Decker's 'deep agnosticism', cf. Decker [5], the agnostic isn't in a position to know one way or another about what, for example, God did or didn't do. In particular, an agnostic realizes that they can't rule out the epistemic possibility of God being the cause of something if they are willing to grant the consistency of God *not* being the cause of that thing. Or to recognize that a proposition may be necessarily so when it is epistemically possible that it is contingently true. If one were to deny that $\langle K \rangle (p\ \&\ \sim \blacksquare p)$, that would mean that $K \sim (p\ \&\ \sim \blacksquare p)$, i.e., one has conclusive evidence for $\sim (p\ \&\ \sim \blacksquare p)$. Similarly for the epistemic possibility of $\blacksquare p$, presuming that p is true. But the agnostic isn't in that position. The agnostic should give fair consideration to both $\blacksquare p$ and $p\ \&\ \sim \blacksquare p$, for any proposition p (save perhaps logical truths).

Definition 2 (Second Grade of Agnostic Involvement). An agent a is *agnostic* (second grade) iff

$$(2A)\ B_a(\forall p)(\langle K_a \rangle (p\ \&\ \sim \blacksquare p) \longleftrightarrow \langle K_a \rangle \blacksquare p)$$

The second grade of involvement maintains that the agent has an attitude toward the absence of knowledge. In this case the agent believes that its knowledge is fair toward—in the sense that it is consistent with—each instance of $\blacksquare p$ and $p\ \&\ \sim \blacksquare p$. This form of agnosticism (and the next) is specific to the particular (forms of) propositions I am considering.

Both the first and second grades may be too strong for some. Some claim that rationality puts *relational* requirements on our beliefs rather than stipulating that there are particular propositions we should believe, cf. Kolodny [9]. Thus, if one is to be rationally fair in a way that is agnostic towards $(\forall p)(p \rightarrow \blacksquare p)$, one needs a form of agnosticism that doesn't require any beliefs. The next form of agnosticism concerns how the agent approaches the relationships between propositions about the consistency of certain propositions with one's epistemic state.

This form of agnostic involvement is weaker than 2A. It can be represented by saying, for example, that one believes that it is consistent with what one knows that God isn't the ultimate cause of some true proposition, if and only if one believes that it is consistent with what one knows that God *is* the cause of that fact. This is the final and third grade of agnostic involvement that I will consider.

Definition 3 (Third Grade of Agnostic Involvement). An agent a is *agnostic* (third grade) iff

$$(3A)\ (\forall p)(B_a \langle K_a \rangle (p\ \&\ \sim \blacksquare p) \longleftrightarrow B_a \langle K_a \rangle \blacksquare p)$$

This condition doesn't seem to require that the agent have a particular attitude. All the agent's beliefs are conditional on having other beliefs about the consistency of $\blacksquare p$ and $p\ \&\ \sim \blacksquare p$ with what the agent knows. According to my discussion in the introduction, an agnostic position should express an attitude, and 3A doesn't seem to do that. But it does express a conditional attitude, or what one might say is a disposition toward beliefs. I think such a disposition is sufficient for a form of agnosticism.

This form of agnosticism might also provide comfort to those who think that S5 really should be the logic of ideal rationality. Since the first grade of agnosticism isn't available to an S5 knower, but we might think that such a knower should still be capable of agnosticism in some cases. The S5 knower has complete command over all that it knows and doesn't know, and belief and knowledge collapse. But still such an agent isn't omniscient. 3A could provide a form of agnosticism for the S5 knower where the proposition at issue is of the form $(\forall p)(p \to \blacksquare p)$. We will see how inappropriate 3A is for that task.

From the rules and axioms laid down in section 2, 3A follows from 2A immediately by $\forall$B and RegB. How do 2A and 3A compare to the standard set by the first grade of agnostic involvement or the intuitive conception? The second grade is stronger than the first grade. For 3A, in a certain sense it is stronger than the first grade, but in another sense, it is weaker.

Both 2A and 3A require that the agent treat the, for example, necessity and contingency of propositions as equal in order for it to describe the state of the agent's beliefs and knowledge. But the conditions make intuitive sense. If one is truly in a state where they do not know a proposition, but they do not know that it is false—in this logical sense of 'know' that I am using—then it makes intuitive sense that contrary propositions should be consistent with one's knowledge. Each of 2A and 3A asks one to put effort into maintaining a fair distribution of beliefs between $p\ \&\ \sim \blacksquare p$ and $\blacksquare p$ *for each proposition* p, rather than just have an attitude toward two particular propositions in the case of A and Ω.

The second grade (2A) is, however, a reflective requirement on one's beliefs. The 2A agnostic *believes* that their beliefs are fair in this way. That is a stronger requirement than that in the first grade since the content of the proposition is more complex and 2A implies A and Ω as we shall see.

But 3A doesn't require that we have any particular beliefs. Because of that, it would seem to be less taxing than the first grade of agnosticism. It requires that there be certain relationships between our beliefs. Our beliefs must be responsive to beliefs about, for example, God's powers in the way 2A says they should be. I would still construe it as a fairness condition, like in the cases of the first and second grades, but what it requires of an agent is more stringent than what is required by A and Ω. Thus it is both stronger in one sense and less stringent in another than the first grade of agnosticism.

The real test of these conceptions of agnosticism is whether they can be compatible with beliefs about $(\forall p)(p \to \blacksquare p)$; that is, can one hold beliefs about, for example, powers of God while remaining agnostic in the three grades? What I intend to show is that agnosticism of any of the kinds about $(\forall p)(p \to \blacksquare p)$ is at odds with having beliefs contrary to that proposition. What I will show in each case is that the form of agnosticism implies some proposition inconsistent with $(\exists p)B(p \,\&\, \sim \blacksquare p)$. Of course, all of this will turn on using the logical assumptions of $L^{\blacksquare}_{KB}$. Mostly what I show is that assumptions which extend $L^{\blacksquare}_{KB}$ are problematic. However, even $L^{\blacksquare}_{KB}$ shows inconsistencies where, intuitively, there shouldn't be. In section 6 I will return to a discussion of the significance of these results.

4 Consequences

The arguments that I will give relate to that in the knowability paradox. In that paradox, it is shown that the assumption that every truth is knowable implies that every truth is known. The sentences used are $(\forall p)(p \to \Diamond Kp)$ and $(\forall p)(p \to Kp)$, respectively. In the knowability proof, one instantiates the knowability formula with the sentence $p \,\&\, \sim Kp$, i.e., that p is true but unknown, then proceeds to show that the consequent of the knowability formula implies a contradiction. But that means the assumption $p \,\&\, \sim Kp$ is false, and so $p \to Kp$ is true. By universal introduction, then, all truths are known.

The analog of p being true but unknown here is $p \,\&\, \sim \blacksquare p$ which can be interpreted as either 'p is the case, but God isn't its truthmaker'; or 'p is the case, but God doesn't know it'; or 'p is contingently true'. For simplicity I will just discuss the consequences in the abstract since I am only using the logical properties for $\blacksquare$ laid out in the last section.

Essentially the same proof can be given in both the case of 3A and 2A. It relies on the following two important and connected facts. First,

Lemma 1. $\blacksquare(p \,\&\, \sim \blacksquare p) \longleftrightarrow \bot$ in $L^{\blacksquare}_{KB}$.

Proof. Right to left follows because contradictions imply everything, and left to right follows from RM$\blacksquare$ and T$\blacksquare$ since RM$\blacksquare$ implies Dist. □

This fact is important because of the role that the formula (p & ~ $\blacksquare p$) plays below. Since it is important, I will give it a special abbreviation: $\text{NCG}(p) =_{df} (p$ & ~ $\blacksquare p)$ (not caused by God). Using the abbreviation and disregarding the agent subscripts, permits writing 2A and 3A in a slightly more readable form.

$$(2A) \ B(\forall p)(\langle K \rangle \text{NCG}(p) \longleftrightarrow \langle K \rangle \blacksquare p)$$

$$(3A) \ (\forall p)(B \langle K \rangle \text{NCG}(p) \longleftrightarrow B \langle K \rangle \blacksquare p)$$

Second, as those familiar with provability logic will know, cf. Verbrugge [21], $\text{NCG}(x)$ has a fixed point property in this logic as a corollary of the first fact. That is, $\text{NCG}(p)$ is a fixed point of $\text{NCG}(x)$, i.e.,

Corollary 1. $\text{NCG}(\text{NCG}(p)) \longleftrightarrow \text{NCG}(p)$ in $L^{\blacksquare}_{KB}$.

I will leave the proof as an exercise (note you only have to use the rules of $\blacksquare$). I will apply these facts to 2A and 3A in a three stage process.

Stage α. Notice that 2A and 3A are each equivalent to conjunctions of two formulas. 2A is equivalent to

$$B(\forall p)(\langle K \rangle \text{NCG}(p) \to \langle K \rangle \blacksquare p) \ \& \ B(\forall p)(\langle K \rangle \blacksquare p \to \langle K \rangle \text{NCG}(p)),$$

and 3A is equivalent to

$$(\forall p)(B \langle K \rangle \text{NCG}(p) \to B \langle K \rangle \blacksquare p) \ \& \ (\forall p)(B \langle K \rangle \blacksquare p \to B \langle K \rangle \text{NCG}(p)).$$

These equivalences are simply seen from the rules of the quantifiers and from the fact that $B(\varphi \ \& \ \psi) \longleftrightarrow (B\varphi \ \& \ B\psi)$ in $L^{\blacksquare}_{KB}$ in the case of 2A. Thus, committing oneself to 2A logically commits one to

$$(2A') \ B(\forall p)(\langle K \rangle \text{NCG}(p) \to \langle K \rangle \blacksquare p),$$

and committing to 3A commits one to

$$(3A')(\forall p)(B \langle K \rangle \text{NCG}(p) \to B \langle K \rangle \blacksquare p)$$

simply by & elimination. It is $2A'$ and $3A'$ which are of interest in what follows.

Stage β. Now we apply the facts from above. From lemma 1,

$$\blacksquare \text{NCG}(p) \longleftrightarrow \bot$$

so using REK, N$\langle K \rangle$ and transitivity, we get

$$\langle K \rangle \blacksquare \text{NCG}(p) \longleftrightarrow \bot. \quad (1)$$

From NecB and DB, it follows that $\neg B\bot$ is a theorem, so, $B\bot \longleftrightarrow \bot$ is as well. Further, applying REK, N$\langle K \rangle$ and RMB gives

$$B\langle K \rangle \bot \longleftrightarrow \bot,$$

hence,

$$B\langle K \rangle \blacksquare \text{NCG}(p) \longleftrightarrow \bot \quad (2)$$

from (1) and the intervening steps. To apply corollary 1 let's look at instantiating

$$(\forall p)(\langle K \rangle \text{NCG}(p) \to \langle K \rangle \blacksquare p)$$

with $\text{NCG}(p)$. We get

$$\langle K \rangle \text{NCG}(\text{NCG}(p)) \to \langle K \rangle \blacksquare \text{NCG}(p).$$

But corollary 1 and REK allow us to derive the equivalent formula

$$\langle K \rangle \text{NCG}(p) \to \langle K \rangle \blacksquare \text{NCG}(p)$$

by replacing $\text{NCG}(\text{NCG}(p))$ in the antecedent with just $\text{NCG}(p)$.

Stage γ. The consequent of $\langle K \rangle \text{NCG}(p) \to \langle K \rangle \blacksquare \text{NCG}(p)$ can be replaced with $\bot$ by (1) which gives

$$\langle K \rangle \text{NCG}(p) \to \bot,$$

so the whole formula is equivalent to: $\sim \langle K \rangle \text{NCG}(p)$. By the duality of $\langle K \rangle$ and K we can get $K \sim \text{NCG}(p)$ which is equivalent to $K(p \to \blacksquare p)$, after replacing $\text{NCG}(p)$ with what it is an abbreviation for and applying some equivalences of propositional logic. Applying TK, $K(p \to \blacksquare p)$ implies $p \to \blacksquare p$. Therefore, we can conclude that

$$(\forall p)(\langle K \rangle \text{NCG}(p) \to \langle K \rangle \blacksquare p) \to (p \to \blacksquare p) \quad (3)$$

is a theorem of $L_{KB}^{\blacksquare}$. We can then apply $\forall I$ to the consequent of the formula above (since p isn't free in the antecedent) and get:

$$(\forall p)(\langle K \rangle \text{NCG}(p) \to \langle K \rangle \blacksquare p) \to (\forall p)(p \to \blacksquare p).$$

Finally, using RMB we can derive that

$$B(\forall p)(\langle K \rangle \operatorname{NCG}(p) \to \langle K \rangle \blacksquare p) \to B(\forall p)(p \to \blacksquare p).$$

That is to say that 2A implies (since 2A' follows from 2A) that the putative agnostic actually *believes* in the omnipotence of God—or is logically committed to it by a belief they have about their epistemic state.

What seemed to be a requirement of epistemic fairness or humility forces the agent to accept something which is much stronger, intuitively. This result, however, shows that a 2A agnostic satisfies condition A since $\varphi \to \langle K \rangle \varphi$ follows from TK, and so $B \langle K \rangle (\forall p)(p \to \blacksquare p)$ follows from 2A by RMB and transitivity of $\to$. So we are guaranteed that 2A provides part of agnosticism in the first grade sense; in fact it also satisfies Ω.[5]

To see this, note the following: since $(\forall p)(p \to \blacksquare p)$ implies $\top \to \blacksquare \top$ by $\forall$E, and $\top$ is a theorem, $(\forall p)(p \to \blacksquare p) \vdash \blacksquare \top$ in $L^\blacksquare_{KB}$. So by RMK and a couple of contrapositions, followed by RMB, we get that $B \langle K \rangle (\forall p)(p \to \blacksquare p) \vdash B \langle K \rangle \blacksquare \top$. Instantiating 3A with $\top$ we would get

$$B \langle K \rangle (\top \;\&\; \sim \blacksquare \top) \longleftrightarrow B \langle K \rangle \blacksquare \top.$$

But $(\top \;\&\; \sim \blacksquare \top) \to (\exists p)(p \;\&\; \sim \blacksquare p)$ is a theorem by $\exists$I. Using RMK, contraposition twice and RMB it follows that $B \langle K \rangle (\top \;\&\; \sim \blacksquare \top) \to B \langle K \rangle (\exists p)(p \;\&\; \sim \blacksquare p)$ is a theorem of $L^\blacksquare_{KB}$. Therefore, A and 3A together imply Ω. Since 2A implies both of A and 3A, 2A implies Ω.

Unfortunately similar problems arise for 3A. We start by instantiating the quantifier in 3A' with $\operatorname{NCG}(p)$ as we did above to get:

$$B \langle K \rangle \operatorname{NCG}(\operatorname{NCG}(p)) \to B \langle K \rangle \blacksquare \operatorname{NCG}(p).$$

Following the same reasoning from above, the consequent can be replaced by $\bot$ using formula (2) while the antecedent is equivalent to $B \langle K \rangle \operatorname{NCG}(p)$, and so the whole formula is equivalent to: $B \langle K \rangle \operatorname{NCG}(p) \to \bot$. This last formula is equivalent to $\sim B \langle K \rangle (p \;\&\; \sim \blacksquare p)$, so by a series of dualities we can push the $\sim$ through the operators to achieve the equivalent formula: $\langle B \rangle K(p \to \blacksquare p)$. Therefore,

$$(\forall p)(B \langle K \rangle \operatorname{NCG}(p) \longleftrightarrow B \langle K \rangle \blacksquare p) \to \langle B \rangle K(p \to \blacksquare p).$$

From here there are a few possible directions of investigation.

First, we can see the logical strength that comes from assuming negative intro-

---

[5] I thank an anonymous referee for pointing this out.

spection. Let's consider the case of NIB. NIB is equivalent to $\langle K \rangle B\varphi \to B\varphi$, from this and KB it follows that $\langle B \rangle B\varphi \to B\varphi$. Since by KB after using RMB and contraposition a couple of times, we can get $\langle B \rangle K\varphi \to \langle B \rangle B\varphi$, therefore we get the corollary that $\langle B \rangle K\varphi \to B\varphi$. And so from $\langle B \rangle K(p \to \blacksquare p)$, applying NIB's corollary and then $\forall I$, we get $(\forall p)B(p \to \blacksquare p)$. An agnostic of the third type would be committed to believing that each true proposition is ultimately God's doing. This is not quite the same as the case of A2, but it is almost as bad and certainly in conflict with suspicions about freewill, if any there be.

Second, there is NI. Note that by the contrapositive of KB, we can derive $\langle K \rangle \varphi$ from $\langle B \rangle \varphi$. So applying this result of KB to $\langle B \rangle K(p \to \blacksquare p)$, we get $\langle K \rangle K(p \to \blacksquare p)$. But again, negative introspection is equivalent to: $\langle K \rangle K\varphi \to K\varphi$. Therefore, we can derive $K(p \to \blacksquare p)$, and by factivity, i.e., TK, and $\forall I$, $(\forall p)(p \to \blacksquare p)$ follows. Thus, negative introspection is incredibly powerful for the third grade agnostic: 3A agnosticism would imply that, for example, all true propositions are necessary—not merely that one must believe it (or that it is a logical consequence of the agent's beliefs).

But both of these facts rely on principles that properly extend $L_{KB}^{\blacksquare}$. What can be said, however, is that 3A$'$ is actually inconsistent with $(\exists p)B(p\ \&\ \sim \blacksquare p)$. That allows us to see that NIB and NI are not necessary to establish that the agnostic cannot harbor *beliefs* about freewill or non-determinism. Since $K\varphi \to \varphi$ is a theorem schema, $K(p \to \blacksquare p) \to (p \to \blacksquare p)$ is a theorem, thus $\langle B \rangle K(p \to \blacksquare p) \to \langle B \rangle (p \to \blacksquare p)$ is also a theorem. Therefore, 3A$'$ (and so 3A) implies $\langle B \rangle (p \to \blacksquare p)$. By $\forall I$, then, 3A$'$ implies $(\forall p)(\langle B \rangle (p \to \blacksquare p))$. But the negation of the latter formula is equivalent to $(\exists p)B(p\ \&\ \sim \blacksquare p)$. Thus, $(\exists p)B(p\ \&\ \sim \blacksquare p)$ is inconsistent with 3A$'$ (and 3A).[6] Mere justified belief can overturn an agnostic of either the 3A or 2A persuasion.

5 What About A?

There is still the first grade of agnosticism to fall back on. It was also the most obvious and intuitive formalization of the position. But it isn't completely innocent either. First let's notice the following:

$$B\langle K \rangle (\forall p)(p \to \blacksquare p) \vdash (\forall p)(Kp \to \langle K \rangle \blacksquare p)\ (*).$$

We can see this as follows. From the (A) condition $B\langle K \rangle (\forall p)(p \to \blacksquare p)$ we can derive, by an application of the contrapositive of $\exists K$ followed by $\forall B$, $(\forall p)B\langle K \rangle (p \to \blacksquare p)$, and instantiating we get $B\langle K \rangle (q \to \blacksquare q)$. By DB and the contrapositive of KB

---

[6]Thank you to an anonymous referee for pointing out this shorter path to inconsistency.

we can derive $\langle K \rangle \langle K \rangle (q \to \blacksquare q)$. By the contrapositive of PI we get $\langle K \rangle (q \to \blacksquare q)$. For any normal modal operator: $\langle K \rangle (\varphi \to \psi) \leftrightarrow (K\varphi \to \langle K \rangle \psi)$, so $(Kq \to \langle K \rangle \blacksquare q)$ follows, and finally by $\forall I$, $(\forall p)(Kp \to \langle K \rangle \blacksquare p)$.

Next, I will consider another principle of a kind I haven't discussed yet since it is an interaction between between $\blacksquare$ and K. It is as follows:

$$(\forall p)(\blacksquare \langle K \rangle p \to \langle K \rangle \blacksquare p) \ [\blacksquare K].$$

I will discuss this principle further in section 6.2. In $L_{KB}^{\blacksquare}$ we can show that the combination of $BK(\blacksquare K)$, i.e., belief that one knows $\blacksquare K$, with A implies 3A′. For that, we need a lemma.

Lemma 2. $B \langle K \rangle \varphi \to BK \langle K \rangle \varphi$ is a valid schema of $L_{KB}^{\blacksquare}$.

Proof.   1. $\langle B \rangle \langle K \rangle \varphi \to \langle K \rangle \langle K \rangle \varphi$ Converse of KB

2. $\langle K \rangle \langle K \rangle \varphi \to \langle K \rangle \varphi$ Converse of PI

3. $\langle B \rangle \langle K \rangle \varphi \to \langle K \rangle \varphi$ By transitivity from 1 and 2

4. $B \langle B \rangle \langle K \rangle \varphi \to \langle B \rangle \langle B \rangle \langle K \rangle \varphi$ Instance of DB

5. $\langle B \rangle \langle B \rangle \langle K \rangle \varphi \to \langle B \rangle \langle K \rangle \varphi$ Instance of the converse of BB

6. $B \langle B \rangle \langle K \rangle \varphi \to \langle B \rangle \langle K \rangle \varphi$ Transitivity from 4 and 5

7. $B \langle B \rangle \langle K \rangle \varphi \to \langle K \rangle \varphi$ Transitivity from 6 and 3

8. $KB \langle B \rangle \langle K \rangle \varphi \to K \langle K \rangle \varphi$ RMK on 7

9. $B \langle B \rangle \langle K \rangle \varphi \to KB \langle B \rangle \langle K \rangle \varphi$ Instance of BKB

10. $B \langle B \rangle \langle K \rangle \varphi \to K \langle K \rangle \varphi$ Transitivity from 8 and 9

11. $B \langle K \rangle \varphi \to BB \langle K \rangle \varphi$ Instance of BB

12. $B \langle K \rangle \varphi \to \langle B \rangle \langle K \rangle \varphi$ Instance of DB

13. $BB \langle K \rangle \varphi \to B \langle B \rangle \langle K \rangle \varphi$ RMB on 12

14. $B \langle K \rangle \varphi \to B \langle B \rangle \langle K \rangle \varphi$ Transitivity from 11 and 13

15. $B \langle K \rangle \varphi \to K \langle K \rangle \varphi$ Transitivity from 10 and 14

16. $BB \langle K \rangle \varphi \to BK \langle K \rangle \varphi$ RMB on 15

17. $B\langle K\rangle\varphi \to BK\langle K\rangle\varphi$ Transitivity from 11 and 16

□

The deduction then proceeds in the following manner:

1. $\vdash (\forall p)(\blacksquare\langle K\rangle p \to \langle K\rangle\blacksquare p) \to (\blacksquare\langle K\rangle(\text{NCG}(t)) \to \langle K\rangle\blacksquare(\text{NCG}(t)))$ by $[\forall E]$ (sub $\text{NCG}(t)$ for p)

2. $\vdash K(\forall p)(\blacksquare\langle K\rangle p \to \langle K\rangle\blacksquare p) \to K(\blacksquare\langle K\rangle(\text{NCG}(t)) \to \langle K\rangle\blacksquare(\text{NCG}(t)))$ by [RMK] on 1

3. $\vdash K(\varphi \to \psi) \to (\langle K\rangle\varphi \to \langle K\rangle\psi)$ for any normal modal operator, so an instance of 3 is

4. $\vdash K(\blacksquare\langle K\rangle(\text{NCG}(t)) \to \langle K\rangle\blacksquare(\text{NCG}(t))) \to (\langle K\rangle\blacksquare\langle K\rangle(\text{NCG}(t)) \to \langle K\rangle\langle K\rangle\blacksquare(\text{NCG}(t)))$, and then

5. $\vdash K(\forall p)(\blacksquare\langle K\rangle p \to \langle K\rangle\blacksquare p) \to (\langle K\rangle\blacksquare\langle K\rangle(\text{NCG}(t)) \to \langle K\rangle\langle K\rangle\blacksquare(\text{NCG}(t)))$ from 2 and 4 by transitivity

6. $\vdash \langle K\rangle\blacksquare(\text{NCG}(t)) \longleftrightarrow \bot$ as we have seen above, and

7. $\vdash \langle K\rangle\bot \longleftrightarrow \bot$ by N$\langle K\rangle$, so

8. $\vdash \langle K\rangle\langle K\rangle\blacksquare(\text{NCG}(t)) \longleftrightarrow \bot$ by REK a few times, and with

9. $\vdash \bot \to \langle K\rangle\blacksquare t$, we can derive that

10. $\vdash (\langle K\rangle\blacksquare\langle K\rangle(\text{NCG}(t)) \to \langle K\rangle\langle K\rangle\blacksquare(\text{NCG}(t))) \to (\langle K\rangle\blacksquare\langle K\rangle(\text{NCG}(t)) \to \langle K\rangle\blacksquare t)$ by PL from 6-9, so

11. $\vdash K(\forall p)(\blacksquare\langle K\rangle p \to \langle K\rangle\blacksquare p) \to (\langle K\rangle\blacksquare\langle K\rangle(\text{NCG}(t)) \to \langle K\rangle\blacksquare t)$ by transitivity from 5 and 10

12. $\vdash B\langle K\rangle(\forall p)(p \to \blacksquare p) \to (K\langle K\rangle\text{NCG}(t) \to \langle K\rangle\blacksquare\langle K\rangle\text{NCG}(t))$ by (*) above and $[\forall E]$, so

13. $\vdash (B\langle K\rangle(\forall p)(p \to \blacksquare p)\ \&\ K(\forall p)(\blacksquare\langle K\rangle p \to \langle K\rangle\blacksquare p)) \to (K\langle K\rangle(\text{NCG}(t)) \to \langle K\rangle\blacksquare t)$ by PL from 11 and 12, hence

14. $\vdash B(B\langle K\rangle(\forall p)(p \to \blacksquare p)\ \&\ K(\forall p)(\blacksquare\langle K\rangle p \to \langle K\rangle\blacksquare p)) \to B(K\langle K\rangle(\text{NCG}(t)) \to \langle K\rangle\blacksquare t)$ by [RMB] on 13.

15. By the normality of B and [BB], $(B\varphi\ \&\ B\psi) \to B(B\varphi\ \&\ \psi)$, thus

16. $\vdash (B \langle K \rangle (\forall p)(p \to \blacksquare p) \,\&\, BK(\forall p)(\blacksquare \langle K \rangle p \to \langle K \rangle \blacksquare p)) \to B(K \langle K \rangle (\text{NCG}(t)) \to \langle K \rangle \blacksquare t)$.

17. $\vdash B(\langle K \rangle \text{NCG}(t) \to \langle K \rangle \blacksquare t) \to (BK \langle K \rangle (\text{NCG}(t)) \to B \langle K \rangle \blacksquare t)$, also because B is normal.

18. By lemma 2, $B \langle K \rangle \text{NCG}(t) \to BK \langle K \rangle \text{NCG}(t)$ is a theorem so,

19. $\vdash B(\langle K \rangle (\text{NCG}(t)) \to \langle K \rangle \blacksquare t) \to (B \langle K \rangle (\text{NCG}(t)) \to B \langle K \rangle \blacksquare t)$, by transitivity from 17 and 18. Hence,

20. $\vdash (B \langle K \rangle (\forall p)(p \to \blacksquare p) \,\&\, BK(\forall p)(\blacksquare \langle K \rangle p \to \langle K \rangle \blacksquare p)) \to (B \langle K \rangle (\text{NCG}(t)) \to B \langle K \rangle \blacksquare t)$ by PL from 17 and 19. Finally,

21. $\vdash (B \langle K \rangle (\forall p)(p \to \blacksquare p) \,\&\, BK(\forall p)(\blacksquare \langle K \rangle p \to \langle K \rangle \blacksquare p)) \to (\forall p)(B \langle K \rangle (\text{NCG}(p)) \to B \langle K \rangle \blacksquare p)$ by $[\forall I]$ on 20.

What this proof shows is that condition A combined with the subjective certainty of $\blacksquare K$ implies 3A′ in $L^{\blacksquare}_{KB}$. As I will argue in section 6.2 believing that one knows $\blacksquare K$ isn't perhaps a stretch for some on some interpretations of $\blacksquare$. Thus, I have made good on my promise to show that each form of agnosticism is incompatible with $(\exists p)B(p \,\&\, {\sim}\blacksquare p)$; although there is some augmentation in the case of A. But it is an extra assumption rather than an extra postulate of epistemic/doxastic logic.

6 Significance

In the following sections I will discuss the interpretations of the $\blacksquare$ operator and whether those interpretations support the logical properties. I will also discuss the acceptability of $\blacksquare K$ on those interpretations. Finally, I will end with comments about what I think should be learned from these results.

6.1 The Interpretations of $\blacksquare$

On the 'God' interpretations of the the $\blacksquare$ operator, the particular propositions I have considered in this paper are whether one can be agnostic about God being the cause of all things (omnipotence), and also God knowing all things (omniscience). The way I could have formalized these claims to make them distinct is as follows:

(OP) $(\forall p)(p \to \blacksquare_P p)$ (omnipotence), and

(OS) $(\forall p)(p \to \blacksquare_K p)$ (omniscience).

In order to represent what is being said when someone says 'God is the cause of all things' I have used an operator ■ₚ. It stands for 'God is the ultimate reason that ...' or 'God is the truthmaker of the proposition that ...' or 'God is the ultimate explanation for the proposition that ...'. Very important caveat: I am not trying to prove the existence of God here. The agnostic in this case is willing to grant some being, but is wondering whether God has any of the powers usually ascribed to God. One may look at Avicenna's argument for the existence of God and say, 'that seems right', but, as many have, been unconvinced by his further derivations of the attributes of God from his definition. What is key is that these readings permit treating this "God" operator as a modal. Thus its attaches to propositions.

The ■_K operator is simply read as 'God knows that ...'. Thus it is a standard, agent relative knowledge operator. The agent just happens to be God.

The intuitive understanding of Dist is that if God is the ultimate cause for A and B, then God is the ultimate cause of A and similarly for B. For RE■, if A is logically equivalent to B, then God is the ultimate cause of A iff God is the ultimate cause of B. Here I am being a bit loose with my terminology since A and B might be considered to be events, but in the logical system φ and ψ must be propositions. However, I don't think there is much to dispute in the principles I have offered so far. In the case of God's knowledge, what this amounts to saying is that God's knowledge in closed under logical consequence. If there was ever a being who was logically omniscient, it is God. Thus, I think we can attribute these properties to ■_K, at least. For the T■ principle, I would argue that for any creature, if it makes it the case that φ, then it is the case that φ. So ■ₚ$\varphi \to \varphi$ should hold. Similarly for the factivity of God's knowledge: ■_K$\varphi \to \varphi$.

On the necessity reading of ■, it is usually assumed to be closed under logical consequence. Also, if φ is necessarily true, it is true. Thus, ■ would conform to RM■ and T■. There are many other possible interpretations that one could give to the ■ operator which might be interesting; van Inwagen's N operator '...is true and no one ever had any influence whether ...' might be of interest, cf. van Inwagen [20]. But I will content myself with just the three above.

Before moving on, I want to consider the case of necessity in more detail. One of the assumptions for modal logics of necessity is that ■⊤ is a theorem; let's call that rule Nec■. That is the standard assumption that logical truths are necessary truths. It follows from that that $B\langle K\rangle$■⊤ would be a theorem of $L_{KB}^{■}$+Nec■. But consider what that means for 3A and 2A. Since $B\langle\top \& \sim ■\top\rangle \leftrightarrow B\langle K\rangle$■⊤ follows from 3A, 3A alone will imply that $B\langle K\rangle(\top \& \sim ■\top)$. But that is equivalent to $B\langle K\rangle(\top \& \bot)$, which is equivalent to $\bot$. Thus, in the case of necessity, 3A and 2A are inconsistent assumptions. They cannot be formulations of agnosticism about determinism in the presence of Nec■. An acceptable revision, is to formulate the second and third grades of agnosticism in terms of 2A′ and 3A′. In those terms, 2A′ will imply 3A′, and 2A′ will imply A, but it does not seem to imply Ω. Nonetheless, it will still hold that $\{BK(\blacksquare K), A, (\exists p)B(p \,\&\, \sim \blacksquare p)\}$, as shown in section 5, is an inconsistent set in $L_{KB}^{\blacksquare}+\text{Nec}\blacksquare$.

6.2 Should One Accept ■K?

The answer depends on the interpretation of $\blacksquare$. I think that ■K *could* be accepted on each of the interpretations, but whether it is more sensible to accept it than reject it is another matter. So what does the principle ■K say? $(\forall p)(\blacksquare \langle K \rangle p \to \langle K \rangle \blacksquare p)$, on the omnipotence reading, says that if God made it so that p is consistent with what I know, then it is consistent with what I know that God made it so that p, for all p. What is the alternative to accepting this principle? Moreover, could I claim to believe to know that it was false? That would require a p such that $\blacksquare \langle K \rangle p$ and $K \sim \blacksquare p$. If I were to claim that, I would have to believe that I knew that God didn't bring about p. That is intuitively at odds with being agnostic about God's omnipotence—in my non-standard sense of that term. At best, however, this would be an argument that ■K is consistent with what I know. Unfortunately the argument above requires that one believe that one knows ■K.

If one believes that God is benevolent, then they might have a reason to reject ■K.[7] If e represents some evil fact, then they would be wont to claim that while God could make it the case that e is consistent with what they know while they would be subjectively certain that God didn't bring about e. It is, they would say, part of the definition of their God that God be all good and so wouldn't bring about that evil. But that kind of person would not be agnostic about God's omnipotence—in the non-standard sense. An agnostic about $(\forall p)(p \to \blacksquare_P p)$ wouldn't hold a view incompatible with ■K.

Turning from reasons to accept ■K, what do we have to accept in accepting ■K? We must accept that God did not make it the case that one doesn't know $(p \to \blacksquare p)$, or more formally $(\forall p) \sim \blacksquare \sim K(p \to \blacksquare p)$. That follows by a similar argument as above, i.e., sub $\text{NCG}(p)$ into ■K: $\blacksquare \langle K \rangle \text{NCG}(p) \to \langle K \rangle \blacksquare \text{NCG}(p)$. So if it is the case that $p \,\&\, \sim \blacksquare p$ is consistent with what I know, then God can't be the cause of that. But all one needs to do to maintain consistency is believe it knows $(\forall p) \sim \blacksquare \sim K(p \to \blacksquare p)$; an agent must believe it knows that God didn't make it the case that it doesn't know $(p \to \blacksquare p)$. While odd, it isn't inconsistent.

My previous discussion has focused on the 'God as cause' operator, and I would like to say a few things about the 'God knows' operator. The arguments above

---

[7] Thank you to an anonymous referee for reminding me of what the benevolence of God would mean.

are all completely formal, so whether their conclusions are acceptable turns on the acceptability of the principles involved. Given the usual presumption of the logical omniscience of God, I would say the principles laid down for ■ are acceptable on the 'knows' reading. Thus, all of the proofs go through. But the question of whether ■K is acceptable is another matter.

Recall ■K, $(\forall p)(\blacksquare\langle K\rangle p \to \langle K\rangle \blacksquare p)$. ■K says that if God knows that p is consistent with what you know, then it is consistent with what you know that God knows that p. For it to be false, there would need to be a p such that God knows that it is consistent with what you know that p, but you know that God doesn't know that p. I ask again: What proposition might be like that? It would seem to be something like $p\ \&\ \sim \blacksquare p$. Indeed, $K \sim \blacksquare(p\ \&\ \sim \blacksquare p)$ is true in the system, but again it is not clear that $\blacksquare\langle K\rangle(p\ \&\ \sim \blacksquare p)$ would be true. The atheist is again in a position where it wouldn't believe $\blacksquare\langle K\rangle(p\ \&\ \sim \blacksquare p)$ for any proposition. But even so, the argument of section 5 doesn't need ■K to be true, just that the agnostic (theist or atheist) *believe* that they know ■K.

For the necessity reading of ■, I think ■K has a better defense. If it is the case that $\blacksquare\langle K\rangle p$, then, by definition, it is necessary that p is consistent with what I know. But that means it cannot be that I can have evidence sufficient to know that $\sim p$. If that is true, then it should be at least consistent with what I know that p is necessary. After all, I can't have evidence against p. Indeed, I could not be in a position where I had evidence against p. For ■K to be false, there would be a proposition p such that $\blacksquare\langle K\rangle p$ and $\sim\langle K\rangle \blacksquare p$. Equivalently, $\sim\blacklozenge K \sim p$ and $K \sim \blacksquare p$: it is impossible for me to know not-p, while I do know that p is not necessarily true. On what basis would I be claiming knowledge that p *could* be false? While not contradictory in $L^{\blacksquare}_{KB}$, it sounds conceptually problematic. Somebody reasoning about necessity should claim to believe that they know this principle on conceptual grounds.

6.3 Final Implications and Interpretation

What we can see from the results of sections 4 and 5 is that any of these forms of agnosticism will result in 3A$'$ (augmented with the $BK(\blacksquare K)$ in the case of the first grade). But one can show that $(\exists p)B(p\ \&\ \sim \blacksquare p)$ is inconsistent with 3A$'$. In fact, one can show that without using even PI as follows.

Let's suppose that t is a witness for $(\exists p)B(p\ \&\ \sim Gp)$, i.e., $B(t\ \&\ \sim Gt)$—or using the abbreviation: $B\text{NCG}(t)$. Since $\varphi \to \langle K\rangle \varphi$ is a theorem, we get $\text{NCG}(t) \to \langle K\rangle \text{NCG}(t)$, and by the normality of B, we can conclude that $B\text{NCG}(t) \to B\langle K\rangle \text{NCG}(t)$ is a theorem of $L^{\blacksquare}_{KB}$.

From corollary 1, $B\text{NCG}(t) \to B\langle K\rangle \text{NCG}(\text{NCG}(t))$ is also a theorem. Since

$B\langle K\rangle G\text{NCG}(t) \longleftrightarrow \bot$ is a theorem, $\sim B\langle K\rangle G\text{NCG}(t) \longleftrightarrow \top$, and so:

$$(B\langle K\rangle \text{NCG}(\text{NCG}(t)) \,\&\, \sim B\langle K\rangle G\text{NCG}(t)) \longleftrightarrow (B\langle K\rangle \text{NCG}(\text{NCG}(t)) \,\&\, \top),$$

but

$$B\langle K\rangle \text{NCG}(\text{NCG}(t)) \longleftrightarrow (B\langle K\rangle \text{NCG}(\text{NCG}(t)) \,\&\, \top),$$

so $B\text{NCG}(t) \to (B\langle K\rangle \text{NCG}(\text{NCG}(t)) \,\&\, \sim B\langle K\rangle G\text{NCG}(t))$ is a theorem. By $\exists I$,

$$B(t \,\&\, \sim Gt) \to (\exists p)(B\langle K\rangle \text{NCG}(p) \,\&\, \sim B\langle K\rangle Gp)$$

is a theorem, and by $\exists E$, since t is not free in the consequent,

$$(\exists p)B(p \,\&\, \sim Gp) \to (\exists p)(B\langle K\rangle \text{NCG}(p) \,\&\, \sim B\langle K\rangle Gp)$$

is a theorem. But the consequent is equivalent to the negation of 3A'. So 3A, which implies 3A' and is implied by 2A, is suspect in a system weaker than $L^{\blacksquare}_{KB}$.

What does all of this mean? Let me offer a recapitulation. My position is that agnosticism should be conceptually compatible with even ideal (theoretical) rationality. But the ability of entertaining hypotheses should also be compatible with theoretical rationality. Perhaps extreme forms of theoretical rationality would require that one suspend *belief* about any proposition for which one lacked conclusive evidence. On the other end of the spectrum, practical or liberal theoretical rationality might demand belief in propositions so that the agent can have motivations to act in important situations. But philosophical investigation seems to be between the two extremes.

A way to model this middle ground between conservative and liberal theoretical rationality is to accept that there is a distinction between $B\varphi$ and $BK\varphi$. An agent may have the former without the latter. But there should also be the possibility of having $BK\varphi$ without $K\varphi$; subjective certainty and knowledge should come apart. Even Stalnaker recognizes the latter distinction. He does not, however, recognize the former. We could say, using the metaphor of entrenchment from belief revision Rott [14], that propositions such that $BK\varphi$ is true are more entrenched than those for which $B\varphi$. Of course when $K\varphi$ is true, φ is among the most entrenched beliefs. Indeed, they are the irrevocable parts of our cognitive commitments since giving them up or changing their status would be a mistake, cf. Segerberg [15].

But what we have seen is that 3A and 2A (or 3A' and 2A') are incompatible with maintaining certain hypotheses which are contrary to $(\forall p)(p \to \blacksquare p)$. But those hypotheses are compatible with the first grade of agnosticism when it is unaugmented. Thus, the obvious conclusion to draw is that 3A and 2A are simply not forms of agnosticism. However, 2A definitely implies the first grade of agnosticism, so it is a stronger form of agnosticism. It must be agnosticism plus something else. It is tempting to think that the something else would be a belief in one's own rational fairness. This is supported by the following fact. First, 2A implies 3A, and then A + 3A proves Ω. Adding 3A to A implies that the negation of $(\forall p)(p \to \blacksquare p)$ is consistent with one's knowledge. So 3A should be the fairness condition. But it isn't clear that 3A + Ω implies A, nor is it clear that $2A$ is equivalent to A + 3A.

The third grade of agnosticism is a deeper commitment to agnosticism than the first grade, in a certain sense. It doesn't require a commitment to any position. The agent needn't even have beliefs about the consistency of $(\forall p)(p \to \blacksquare p)$, only dispositions to believe certain consistency claims. That is the point of using a (bi)conditional to state the condition. The 3A agnostic is, however, more agnostic than the first grade agnostic. A 3A agnostic can't have beliefs contrary to $(\forall p)(p \to \blacksquare p)$. For that reason it might seem that 3A agnosticism *is* a form of agnosticism that an S5 ideally rational agent could take. But we can see that that too is unsatisfactory since such an agnostic could prove that $(\forall p)(p \to \blacksquare p)$ was, in fact, *true*. Thus, in the end, if I have properly captured rational fairness (for $(\forall p)(p \to \blacksquare p)$) with 3A, then an individual that can take $(\forall p)(p \to \blacksquare p)$ seriously cannot even have good reason to believe that $(\forall p)(p \to \blacksquare p)$ is false.

Are there ways to avoid these conclusions? Of course. One obvious way is to disallow doxastic/epistemic modalities to fall within the scope of other similar modalities. That would block all of the proofs in sections 4 and 5. It would also force rejection of axioms like PI. But part of epistemic game theory, for example, is the ability to reason about what I know that you believe that I know. The ability to iterate modalities of these kinds has become essential. In so far as we want to put our theories of ideal rationality into reflective equilibrium, iteration should be acceptable. But what we could be seeing is a deeper problem having to do with treating epistemic and doxastic logic as *logic* at all. Perhaps what the real solution should be is to remove assumptions about closure under logical consequence all together (i.e., do away with rules of the RMK and NK form). Those rules force logical relations to hold between propositions whose primary operators are not usually considered logical. Since Etchemendy [6, 7], we have recognized that what counts as a logical operator is something that is more a matter of choice than handed down from on high. Various model-theoretic accounts of what operators are logical have been offered, starting with Corcoran and Tarski [4] and extended by others, e.g., Sher [16], but it isn't clear that these programs are unproblematic. What one could suggest is that the operators are not logical precisely because these unwelcome connections hold. There are certain relationships of consistency and consequence that are sacrosanct; if a modal logic runs afoul of those, its operators should be suspect. Then again, maybe that kind of idea is simply an expression of logical dogma.

Dogma aside, it is difficult to say what *the* solution to problems like the paradox of the knower, the Church-Fitch Paradox, and that of the above should be. Excluding modal logical formulas from being *logical* axioms may be too strong an approach. Perhaps only a piecemeal, case by case approach to these kinds of problems is called for. The guiding principle in epistemic logic seems to be that we do not want it to tell us what philosophical views we can and cannot hold. As long as a logic doesn't show a philosophical bias, then the logic will be acceptable.

References

[1] G. Aldo Antonelli and Richmond H. Thomason. Representability in second-order propositional poly-modal logic. *The Journal of Symbolic Logic*, 67(3): 1039–54, 2002.

[2] Robert Aumann and Adam Brandenburger. Epistemic conditions for nash equilibrium. *Econometrica*, 63(5):pp. 1161–1180, 1995. ISSN 00129682. URL http://www.jstor.org/stable/2171725.

[3] R.A. Bull. On modal logic with propositional quantifiers. *The Journal of Symbolic Logic*, 34(2):257–63, 1969.

[4] John Corcoran and Alfred Tarski. What are logical notions? *History and Philosophy of Logic*, 7(2):143–154, 1986.

[5] Jason Decker. Disagreement, evidence, and agnosticism. *Synthese*, 187(2):753–783, 2012.

[6] John Etchemendy. *The Concept of Logical Consequence*. Harvard University Press, Cambridge, MA, 1990.

[7] John Etchemendy. Reflections on consequence. In Douglas. Patterson, editor, *New Essays on Tarski and Philosophy*, chapter 11, pages 263–99. Oxford University Press, Oxford, 2008.

[8] Jaakko Hintikka. *Knowledge and Belief.* Ithaca, N.Y.,Cornell University Press, 1962.

[9] Niko Kolodny. Why be rational? *Mind*, 114(455):509–563, 2005.

[10] Thomas S. Kuhn. *Objectivity, Value Judgment, and Theory Choice*, chapter 13, pages 320–39. University of Chicago Press, Chicago and London, 1977.

[11] Robin Le Poidevin. *Agnosticism: A Very Short Introduction*. OUP Oxford, 2010.

[12] Louis P. Pojman. Agnosticism. In Robert Audi, editor, *The Cambridge Dictionary of Philosophy*. Cambridge University Press, 1999.

[13] Sven Rosenkranz. Agnosticism as a third stance. *Mind*, 116(461):55–104, 2007.

[14] Hans Rott. Preferential belief change using generalized epistemic entrenchment. *Journal of Logic, Language and Information*, 1(1):45–78, 1992.

[15] Krister Segerberg. Irrevocable belief revision in dynamic doxastic logic. *Notre Dame Journal of Formal Logic*, 39(3):287–306, 1998.

[16] Gila Sher. *The Bounds of Logic: A Generalized Viewpoint*. MIT Press, Boston, 1991.

[17] Craig Smorynski. The incompleteness theorems. In Jon Barwise, editor, *Handbook of Mathematical Logic*, pages 821–865. North-Holland, 1977.

[18] Robert Stalnaker. On Logics of Knowledge and Belief. *Philosophical Studies*, 128(1):169–199, 2006.

[19] Johannes Stern and Martin Fischer. Paradoxes of interaction? *Journal of Philosophical Logic*, 44(3):287–308, 2015. ISSN 0022-3611. doi: 10.1007/s10992-014-9319-5. URL http://dx.doi.org/10.1007/s10992-014-9319-5.

[20] Peter van Inwagen. The incompatibility of free will and determinism. *Philosophical Studies*, 27(March):185–99, 1975.

[21] Rineke (L.C.) Verbrugge. Provability logic. In Edward N. Zalta, editor, *The Stanford Encyclopedia of Philosophy*. Summer 2014 edition, 2014.

Intuitionistic Modal Logic Made Explicit

Michel Marti*
Institute of Computer Science, University of Bern
mmarti@inf.unibe.ch

Thomas Studer
Institute of Computer Science, University of Bern
tstuder@inf.unibe.ch

Abstract

The logic of proofs of Heyting arithmetic includes explicit justifications for all admissible rules of intuitionistic logic in order to satisfy completeness with respect to provability semantics. We study the justification logic iJT4, which does not have these additional justification terms. We establish that iJT4 is complete with respect to modular models, which provide a Kripke-style semantics, and that there is a realization of intuitionistic S4 into iJT4. Hence iJT4 can be seen as an explicit version of intuitionistic S4.

1 Introduction

Justification logics are explicit modal logics in the sense that they unfold the □-modality in families of so-called justification terms. Instead of formulas □A, meaning that A is known, justification logics include formulas $t : A$, meaning that A is known for reason t.

The original semantics for the first justification logic, the Logic of Proofs LP, was Artemov's provability semantics that interpreted $t : A$ roughly as *t represents a proof of A* in the sense of a formal proof predicate in Peano Arithmetic [1, 2, 21].

Later Fitting [15] interpreted justifications as evidence in a more general sense and introduced epistemic, i.e., possible world, models for justification logics. These models have been further developed to modular models as we use them in this paper [6, 19]. This general reading of justification led to many applications in epistemic logic [4, 5, 8, 9, 10, 11, 12, 16, 18, 20].

We would like to thank the anonymous referees for many helpful comments.
*Supported by the SNSF grant 200020_153169 *Structural Proof Theory and the Logic of Proofs*

Given the interpretation of LP in Peano Arithmetic, it was a natural question to find an intuitionistic version iLP of LP that is the logic of proofs of Heyting Arithmetic. The work by Artemov and Iemhoff [7] and later by Dashkov [13] provides such an iLP. It turned out that iLP is not only LP with the underlying logic changed to intuitionistic propositional logic. In order to get a complete axiomatization with respect to provability semantics, one also has to include certain admissible rules of Heyting Arithmetic as axioms in iLP so that they are represented by novel proof terms.

The main question of this paper is *what is the justification counterpart of intuitionistic* S4. We find that the additional axioms of iLP are not needed if we are interested in completeness with respect to possible world models. We study the intuitionistic justification logic iJT4, which is simply LP over an intuitionistic base instead of a classical one but without any additional axioms. We introduce modular models for iJT4 that are inspired by the Kripke-style semantics for intuitionistic S4 and establish completeness of iJT4 with respect to these models.

Artemov [3] already considerend iJT4, under the name $\mathcal{ILP}$, to provide a provability interpretation of modal λ-terms. In order to achieve this, he established that there is a realization of iS4 into iJT4. We will restate that result as it shows that the justification logic iJT4 indeed is the explicit version of the intuitionistic modal logic iS4.

2 Intuitionistic Modal Logic

We present the intuitionistic modal logic iS4. We will start with introducing the language $\mathcal{L}_I$ of iS4. For our purpose, we will only consider the $\Box$-modality but not the $\Diamond$-modality.

Definition 2.1 (Intuitionistic modal language). We assume a countable set Prop of atomic propositions. The set of formulas $\mathcal{L}_I$ is inductively defined by:

1. every atomic proposition is a formula;

2. the constant symbol $\bot$ is a formula;

3. If A and B are formulas, then $(A \wedge B)$, $(A \vee B)$ and $(A \to B)$ are formulas;

4. if A is a formula, then $\Box A$ is a formula.

There are various semantics available for intuitionistic modal logic. The intuitionistic Kripke models that we introduce in this section are the same as Ono's [22] I-models of type 0. Moreover, these models are equivalent to the models used by

Fischer Servi [14], Plotkin and Stirling [23], and Simpson [24]. Jäger and Marti [17] provide a detailed discussion and comparison of these different approaches.

The semantics for iS4 is given by Kripke models that use two accessibility relations: $\leq$ to model the intuitionistic base logic and R to interpret the $\Box$-modality.

Definition 2.2. An *intuitionistic Kripke model* for iS4 is a tuple

$$\mathfrak{M} = (W, \leq, R, V)$$

such that

(i) $W \neq \varnothing$

(ii) R is a reflexive and transitive binary relation on W

(iii) $\leq$ is a partial order (reflexive and transitive) on W

(iv) $V : \mathsf{Prop} \to \mathcal{P}(W)$, and for any atomic proposition p, the set $V(p)$ is upwards closed, i.e., : $w \leq v, w \in V(p) \implies v \in V(p)$

(v) $w \leq v \implies R[v] \subseteq R[w]$

where $R[v] := \{w \in W \mid (v, w) \in R\}$.

Usually, the definition of intuitionistic Kripke models does not include Condition (ii). Since we exclusively work with Kripke models for intuitionistic S4, we make it part of our definition in order to have a simpler terminology.

Definition 2.3 (Satisfaction in Kripke models). We define the *satisfaction relation* $(\mathfrak{M}, w) \vDash A$ by induction on the $\mathcal{L}_\mathsf{I}$-formula A.

- $(\mathfrak{M}, w) \nvDash \bot$;

- $(\mathfrak{M}, w) \vDash p$ iff $w \in V(p)$;

- $(\mathfrak{M}, w) \vDash A \wedge B$ iff $(\mathfrak{M}, w) \vDash A$ and $(\mathfrak{M}, w) \vDash B$;

- $(\mathfrak{M}, w) \vDash A \vee B$ iff $(\mathfrak{M}, w) \vDash A$ or $(\mathfrak{M}, w) \vDash B$;

- $(\mathfrak{M}, w) \vDash A \to B$ iff $(\mathfrak{M}, v) \vDash B$ for all $v \geq w$ with $(\mathfrak{M}, v) \vDash A$;

- $(\mathfrak{M}, w) \vDash \Box A$ iff $(\mathfrak{M}, v) \vDash A$ for all $v \in R[w]$.

An $\mathcal{L}_\mathsf{I}$-formula A is *valid with respect to Kripke models* if for all Kripke models $\mathfrak{M} = (W, \leq, R, V)$ and all $w \in W$ we have $(\mathfrak{M}, w) \vDash A$.

Intuitionistic modal logic has the following monotonicity property.

Lemma 2.4 (Monotonicity).
$$(\mathfrak{M}, w) \vDash A \text{ and } w \leq v \implies (\mathfrak{M}, v) \vDash A.$$

We will need two different deductive systems for iS4. The system HiS4 is a Hilbert-style calculus whereas the system GiS4 is a Gentzen-style sequent calculus for the intuitionistic modal logic iS4.

Definition 2.5 (The proof system HiS4). The system HiS4 consists of the following axioms:

- All axioms for intuitionistic propositional logic
- $\Box(A \to B) \to (\Box A \to \Box B)$ (K)
- $\Box A \to A$ (T)
- $\Box A \to \Box\Box A$ (4)

The rules of HiS4 are modus ponens and necessitation:

$$\frac{A \to B \quad A}{B} \text{ (MP)} \qquad \frac{A}{\Box A} \text{ (nec)}$$

Theorem 2.6. HiS4 is sound and complete with respect to intuitionistic Kripke models.

Proof. Soundness and completeness follow from [22, Theorem 3.2 on p. 696] and the observation that that Ono's I-models of type 0 are the same as our intuitionistic Kripke models. □

Definition 2.7 (The proof system GiS4). A *sequent* is an expression of the form $\Gamma \supset A$, where Γ is a finite multiset of formulas and A is a formula. The Gentzen-style system GiS4 derives sequents and consists of the following axioms and rules:

$$\Gamma \supset A \quad \text{if } A \in \Gamma \text{ or } \bot \in \Gamma$$

$$\frac{\Gamma, A \supset C \quad \Gamma, B \supset C}{\Gamma, A \vee B \supset C} \; (\vee \supset)$$

$$\frac{\Gamma \supset A}{\Gamma \supset A \vee B} \; (\supset \vee)_1 \qquad \frac{\Gamma \supset B}{\Gamma \supset A \vee B} \; (\supset \vee)_2$$

$$\frac{\Gamma, A, B \supset C}{\Gamma, A \wedge B \supset C} (\wedge \supset) \qquad \frac{\Gamma \supset A \quad \Gamma \supset B}{\Gamma \supset A \wedge B} (\supset \wedge)$$

$$\frac{\Gamma \supset A \quad \Gamma, B \supset C}{\Gamma, A \to B \supset C} (\to \supset) \qquad \frac{\Gamma, A \supset B}{\Gamma \supset A \to B} (\supset \to)$$

$$\frac{A, \Gamma \supset B}{\Box A, \Gamma \supset B} (\Box \supset) \qquad \frac{\Box \Gamma \supset A}{\Box \Gamma \supset \Box A} (\supset \Box)$$

$$\frac{\Gamma \supset A}{\Gamma, \Delta \supset A} \text{ (weakening)} \qquad \frac{\Gamma, A, A \supset B}{\Gamma, A \supset B} \text{ (contraction)}$$

In the rule $(\supset \Box)$, the expression $\Box\Gamma$ denotes the multiset $\{\Box A \mid A \in \Gamma\}$. As usual, we say that a formula A is provable in GiS4, in symbols $\vdash_{\mathsf{GiS4}} A$, if the sequent $\supset A$ is provable.

Theorem 2.8. GiS4 is sound and complete with respect to intuitionistic Kripke models.

Proof. In this proof, theorems and systems refer to [22]. By Theorem 3.2 the Hilbert system L_0 is complete with respect to I-models of type 0, which are the same as our intuitionistic Kripke models. By Theorem 2.1, the sequent system G_0 is equivalent to the Hilbert system L_0. The sequent systems G_0 and K_0 are equivalent, and by Theorem 3.3, we have cut-elimination for K_0. Therefore, the cut-free version of K_0 is complete with respect to intuitionistic Kripke models. Finally observe that cut-free K_0 is the same as our GiS4. □

3 Intuitionistic Justification Logic

In this section, we introduce the syntax for the justification logic iJT4$_{\mathsf{CS}}$, which is the explicit analogue of the intuitionistic modal logic iS4.

Definition 3.1 (Justification terms). We assume a countable set of justification constants and a countable set of justification variables. The set of justification terms Tm is inductively defined by:

1. each justification constant and each justification variable is a justification term;

2. if s and t are justification terms, then so are

- $(s \cdot t)$, read s dot t,
- $(s + t)$, read s plus t,
- $!s$, read bang s.

Definition 3.2 (Formulas). We start with the same set Prop of atomic propositions as in $\mathcal{L}_I$. The set of formulas $\mathcal{L}_J$ is inductively defined by:

1. every atomic proposition is a formula;
2. the constant symbol $\bot$ is a formula;
3. If A and B are formulas, then $(A \wedge B)$, $(A \vee B)$ and $(A \rightarrow B)$ are formulas;
4. if A is a formula and t a term, then $t : A$ is a formula.

Definition 3.3. The axioms of iJT4 consist of the following axioms:

1. all axioms for intuitionistic propositional logic
2. $t : (A \rightarrow B) \rightarrow (s : A \rightarrow t \cdot s : B)$
3. $t : A \rightarrow t + s : A$ and $s : A \rightarrow t + s : A$
4. $t : A \rightarrow A$
5. $t : A \rightarrow !t : t : A$

A *constant specification* CS is any subset

$$\mathsf{CS} \subseteq \{(c, A) \mid c \text{ is a constant and } A \text{ is an axiom of iJT4}\}.$$

A constant specification CS is called:

- *axiomatically appropriate* if for each axiom A of iJT4, there is a constant c such that $(c, A) \in \mathsf{CS}$.

- *schematic* if for each constant c, the set of axioms $\{A \mid (c, A) \in \mathsf{CS}\}$ consists of all instances of several (possibly zero) axiom schemes of iJT4.

For a constant specification CS the deductive system iJT4$_{\mathsf{CS}}$ is the Hilbert system given by the axioms above and by the rules modus ponens and axiom necessitation:

$$\dfrac{A \rightarrow B \quad A}{B} \text{ (MP)} \qquad \dfrac{(c, A) \in \mathsf{CS}}{c : A} \text{ (AN)}$$

Remark 3.4. Although axiom necessitation is a rule without premises, it is important to consider it as a rule and not as an axiom schema. If we said that $c : A$ is an axiom for each $(c, A) \in \mathsf{CS}$, then the notion of an axiom would depend on the constant specification, which in turn would depend on the notion of an axiom. Since we want to avoid this circularity, axiom necessitation is introduced as a rule.

Remark 3.5. Let Tot be the total constant specification, i.e.

$$\mathsf{Tot} := \{(c, A) \mid c \text{ is a constant and } A \text{ is an axiom of iJT4}\}.$$

Artemov's [3] intuitionistic logic of proofs $\mathcal{ILP}$ is then the same as our $\mathsf{iJT4}_{\mathsf{Tot}}$.

As usual in justification logic, we can establish the Deduction Theorem and the Lifting Lemma.

Theorem 3.6 (Deduction Theorem). *For every set of formulas M and all formulas A, B we have that*

$$M \cup \{A\} \vdash_{\mathsf{iJT4}_{\mathsf{CS}}} B \iff M \vdash_{\mathsf{iJT4}_{\mathsf{CS}}} A \to B.$$

Lemma 3.7 (Lifting Lemma). *Let CS be an axiomatically appropriate constant specification. For arbitrary formulas $A, B_1, \ldots, B_m, C_1, \ldots, C_n$ and arbitrary justification terms $r_1, \ldots, r_m, s_1, \ldots, s_n$, if*

$$r_1 : B_1, \ldots, r_m : B_m, C_1, \ldots, C_n \vdash_{\mathsf{iJT4}_{\mathsf{CS}}} A,$$

then there is a justification term t such that

$$r_1 : B_1, \ldots, r_m : B_m, s_1 : C_1, \ldots, s_n : C_n \vdash_{\mathsf{iJT4}_{\mathsf{CS}}} t : A.$$

Definition 3.8 (Substitution). A *substitution* is a mapping from justification variables to justification terms. Given a substitution σ and an $\mathcal{L}_\mathsf{J}$-formula A, the formula $A\sigma$ is obtained from A by simultaneously replacing all occurrences of x with $\sigma(x)$ in A for all justification variables x.

As usual in justification logic, we have the following substitution property for schematic constant specifications.

Lemma 3.9 (Substitution Property). *Let CS be a schematic constant specification. We have for any $\mathcal{L}_\mathsf{J}$-formula A and any substitution σ*

$$B_1, \ldots, B_n \vdash_{\mathsf{iJT4}_{\mathsf{CS}}} A \quad \text{implies} \quad B_1\sigma, \ldots, B_n\sigma \vdash_{\mathsf{iJT4}_{\mathsf{CS}}} A\sigma.$$

We find that $\mathsf{iJT4}_{\mathsf{CS}}$ is a conservative extension of intuitionistic propositional logic. Hence $\mathsf{iJT4}_{\mathsf{CS}}$ is consistent.

Lemma 3.10 (Conservativity). iJT4$_{CS}$ is a conservative extension of intuitionistic propositional logic Int, i.e., for any formula A of intuitionistic propositional logic,

$$\vdash_{\mathsf{iJT4_{CS}}} A \quad \text{iff} \quad \vdash_{\mathsf{Int}} A.$$

Proof. The implication from right to left is trivial. For the other direction consider the mapping $(\cdot)^s$ from $\mathcal{L}_J$ to formulas of intuitionistic propositional logic given by:

$$\bot^s := \bot \qquad\qquad p^s := p$$
$$(A \wedge B)^s := A^s \wedge B^s \qquad (A \vee B)^s := A^s \vee B^s$$
$$(A \to B)^s := A^s \to B^s \qquad (t : B)^s := B^s$$

For any formula C of $\mathcal{L}_J$, we can show

$$\vdash_{\mathsf{iJT4_{CS}}} C \quad \text{implies} \quad \vdash_{\mathsf{Int}} C^s$$

by induction on the length of the iJT4$_{CS}$-derivation. Thus the claim immediately follows from $A^s = A$. □

Lemma 3.11 (Consistency of iJT4$_{CS}$). For any constant specification CS, the logic iJT4$_{CS}$ is consistent.

Proof. Assume towards a contradiction that iJT4$_{CS}$ were not consistent, that means $\vdash_{\mathsf{iJT4_{CS}}} \bot$. By the conservativity of iJT4$_{CS}$ over propositional intuitionistic logic Int (previous lemma), it would then follow that $\vdash_{\mathsf{Int}} \bot$, which is not the case. □

4 Basic Modular Models

Basic modular models are syntactic models for justification logic. Yet, our basic modular models will include possible worlds in order to deal with the intuitionistic base logic. After defining basic modular models for intuitionistic justification logic, we will prove soundness and completeness.

In this and the next section, derivability always refers to derivability in iJT4$_{CS}$. Accordingly we use $\vdash$ to mean $\vdash_{\mathsf{iJT4_{CS}}}$.

For two sets of formulas S, T and a term s we write

$$S \cdot T := \{F \mid G \to F \in S \text{ and } G \in T \text{ for some formula } G\}$$
$$s : S := \{s : F \mid F \in S\}$$

Definition 4.1 (Basic evaluation). A *basic evaluation* is a tuple $(W, \leq, *)$ where

$$W \neq \emptyset \text{ and } \leq \text{ is a partial order on } W,$$

$$* : \mathsf{Prop} \times W \to \{0, 1\} \qquad * : \mathsf{Tm} \times W \to \mathcal{P}(\mathcal{L}_\mathsf{J})$$

(where we often write t_w^* for $*(t, w)$ and p_w^* for $*(p, w)$), such that for arbitrary $s, t \in \mathsf{Tm}$, any formula A, and every $w \in W$,

(1) $s_w^* \cdot t_w^* \subseteq (s \cdot t)_w^*$;

(2) $s_w^* \cup t_w^* \subseteq (s + t)_w^*$;

(3) $(t, A) \in \mathsf{CS} \implies A \in t_w^*$;

(4) $s : s_w^* \subseteq (!s)_w^*$.

Furthermore, it has to satisfy the following monotonicity conditions:

(M1) $p_w^* = 1$ and $w \leq v \implies p_v^* = 1$;

(M2) $w \leq v \implies t_w^* \subseteq t_v^*$.

Strictly speaking we should use the notion of a CS basic evaluation because condition (3) depends on a given CS. However, the constant specification will always be clear from the context and we can safely omit it. The same also holds for modular models (to be introduced later).

Definition 4.2 (Truth under basic evaluation). Let $\mathfrak{M} = (W, \leq, *)$ be a basic evaluation. For $w \in W$, we define $(\mathfrak{M}, w) \vDash A$ by induction on the formula A as follows:

- $(\mathfrak{M}, w) \nvDash \bot$;

- $(\mathfrak{M}, w) \vDash p$ iff $p_w^* = 1$;

- $(\mathfrak{M}, w) \vDash A \wedge B$ iff $(\mathfrak{M}, w) \vDash A$ and $(\mathfrak{M}, w) \vDash B$;

- $(\mathfrak{M}, w) \vDash A \vee B$ iff $(\mathfrak{M}, w) \vDash A$ or $(\mathfrak{M}, w) \vDash B$;

- $(\mathfrak{M}, w) \vDash A \to B$ iff $(\mathfrak{M}, v) \vDash B$ for all $v \geq w$ with $(\mathfrak{M}, v) \vDash A$;

- $(\mathfrak{M}, w) \vDash t : A$ iff $A \in t_w^*$.

We immediately obtain the monotonicity property for intuitionistic justification logic.

Lemma 4.3 (Monotonicity). For any basic evaluation $\mathfrak{M} = (W, \leq, *)$, any $w, v \in W$, and any formula A:

$$(\mathfrak{M}, w) \vDash A \text{ and } w \leq v \implies (\mathfrak{M}, v) \vDash A.$$

Definition 4.4 (Factive evaluation). A basic evaluation $\mathfrak{M} = (W, \leq, *)$ is called *factive* iff

$$A \in t_w^* \implies (\mathfrak{M}, w) \vDash A$$

for all formulas A, all justification terms t and all states $w \in W$.

Definition 4.5 (Basic modular model). A *basic modular model* is a basic evaluation $(W, \leq, *)$ that is factive.

We say that a formula A is *valid with respect to basic modular models* if for all basic modular models $\mathfrak{M} = (W, \leq, *)$ and all $w \in W$ we have $(\mathfrak{M}, w) \vDash A$.

Lemma 4.6 (Soundness of iJT4$_{CS}$ with respect to basic modular models). For every formula A:

$$\vdash A \quad \text{implies} \quad A \text{ is valid with respect to basic modular models.}$$

In order to show completeness, we need some auxiliary definitions and lemmas.

Definition 4.7. We call a set of formulas Δ *prime* iff it satisfies the following conditions:

(i) Δ has the disjunction property, i.e., $A \vee B \in \Delta \implies A \in \Delta$ or $B \in \Delta$;

(ii) Δ is deductively closed, i.e., for any formula A, if $\Delta \vdash A$, then $A \in \Delta$;

(iii) Δ is consistent, i.e., $\bot \notin \Delta$.

From now on, we will use Σ, Δ, Γ for prime sets of formulas.

Lemma 4.8. Let N be an arbitrary set of formulas and let A, B and C be formulas. If

$$N \cup \{A \vee B\} \nvdash C, \text{ then } N \cup \{A\} \nvdash C \text{ or } N \cup \{B\} \nvdash C.$$

Proof. By contraposition. Assume that

$$N \cup \{A\} \vdash C \text{ and } N \cup \{B\} \vdash C$$

Then there are finite subsets $N_1 \subseteq N \cup \{A\}$ and $N_2 \subseteq N \cup \{B\}$ such that

$$N_1 \vdash C \text{ and } N_2 \vdash C$$

Now let $N'_1 := N_1 \setminus \{A\}$ and $N'_2 := N_2 \setminus \{B\}$. Then N'_1, N'_2 are finite subsets of N, and
$$(N'_1 \cup \{A\}) \vdash C \text{ and } (N'_2 \cup \{B\}) \vdash C$$
So by the Deduction Theorem,
$$N'_1 \vdash A \to C \text{ and } N'_2 \vdash B \to C$$
So
$$N'_1 \vdash (A \to C) \text{ and } N'_2 \vdash (B \to C).$$
Strengthening the antecedent, we get
$$(N'_1 \cup N'_2) \vdash (A \to C) \text{ and } (N'_1 \cup N'_2) \vdash (B \to C)$$
and, therefore,
$$(N'_1 \cup N'_2) \vdash ((A \to C) \land (B \to C)).$$
By propositional reasoning we get
$$(N'_1 \cup N'_2) \vdash ((A \lor B) \to C),$$
By the Deduction Theorem it follows that
$$(N'_1 \cup N'_2 \cup \{A \lor B\}) \vdash C.$$
Since N'_1 and N'_2 are finite subsets of N, $N'_1 \cup N'_2 \cup \{A \lor B\}$ is a finite subset of $N \cup \{A \lor B\}$, so by definition
$$N \cup \{A \lor B\} \vdash C. \qquad \square$$

Theorem 4.9 (Prime Lemma). *Let B be a formula and let N be a set of formulas such that $N \nvdash B$. Then there exists a prime set Π with $N \subseteq \Pi$ and $\Pi \nvdash B$.*

Proof. Let $(A_n)_{n \in \mathbb{N}}$ be an enumeration of all formulas.
    Now we define $N_0 := N$,
$$N_{i+1} := \begin{cases} N_i \cup \{A_i\}, & \text{if } N_i \cup \{A_i\} \nvdash B \\ N_i, & \text{otherwise} \end{cases}$$
and finally
$$N^\star := \bigcup_{i \in \mathbb{N}} N_i$$
By induction in i, one can easily show that for all $i \in \mathbb{N} : N_i \nvdash B$ and, therefore, $N^\star \nvdash B$.
    It remains to show that $N^\star$ is prime. We have the following:

- $\bot \notin N^\star$: We have $N^\star \nvdash B$, hence $\bot \notin N^\star$.

- $N^\star$ is deductively closed: Assume it is not, i.e., there is a formula A with
$$N^\star \vdash A \text{ but } A \notin N^\star$$

Since $N^\star \vdash A$ but $N^\star \nvdash B$, we know that
$$N^\star \cup \{A\} \nvdash B$$

Otherwise, by the Deduction Theorem 3.6
$$N^\star \vdash A \to B \text{ and } N^\star \vdash A$$

so by propositional reasoning,
$$N^\star \vdash B, \text{ which contradicts our observation above.}$$

Since $(A_n)_{n \in \mathbb{N}}$ is an enumeration of all formulas, there is some i such that $A = A_i$. But then
$$N_i \cup \{A_i\} \nvdash B.$$

So by construction
$$N_{i+1} = N_i \cup \{A_i\}$$

and, therefore,
$$A = A_i \in N_{i+1} \subseteq N^\star,$$

which contradicts our assumption.

- $N^\star$ has the disjunction property: Assume that $C \vee D \in N^\star$. Then there is some i such that $C \vee D = A_i$ and there are i_1, i_2 such that
$$C = A_{i_1} \text{ and } D = A_{i_2}$$

Now we have
$$N^\star = N^\star \cup \{C \vee D\} \nvdash B$$

By the lemma above it follows that
$$N^\star \cup \{C\} \nvdash B \text{ or } N^\star \cup \{D\} \nvdash B$$

In the first case, we have that

$$N_{i_1} \cup \{A_{i_1}\} \nvdash B$$

so by the definition of N_{i_1+1},

$$N_{i_1+1} = N_{i_1} \cup \{A_{i_1}\} = N_{i_1} \cup \{C\}$$

which means that $C \in N_{i_1+1}$ and therefore $C \in N^\star$. The second case is analogous. □

Lemma 4.10. Let Δ be a prime set and t be a justification term. Then

$$t^{-1}\Delta := \{A \mid t : A \in \Delta\} \subseteq \Delta.$$

Proof. Let $A \in t^{-1}\Delta$. Then $t : A \in \Delta$. Since Δ is deductively closed, it contains all axioms, thus $t : A \to A \in \Delta$. Again, since Δ is deductively closed, it follows by (MP) that $A \in \Delta$. □

Definition 4.11 (Canonical basic modular model). The canonical basic modular model is

$$B^{can} := (W^{can}, \leq^{can}, *^{can})$$

where

(i) $W^{can} := \{\Delta \subseteq \mathcal{L}_J \mid \Delta \text{ is prime}\}$

(ii) $\leq^{can} := \subseteq$

(iii) $*^{can}(p, \Delta) = 1$ iff $p \in \Delta$

(iv) $*^{can}(t, \Delta) := t^{-1}\Delta$

Lemma 4.12. B^{can} is a basic evaluation.

Proof. $W \neq \varnothing$: By the consistency of iJT4$_{\mathsf{CS}}$ we have that $\varnothing \nvdash \bot$, it follows by the Prime Lemma 4.9 that there exists a prime set, so $W^{can} \neq \varnothing$.

Next, we check the conditions on the sets of formulas t_Δ^{*can}.

(1) $s_\Delta^{*can} \cdot t_\Delta^{*can} \subseteq (s \cdot t)_\Delta^{*can}$. Let $A \in s_\Delta^{*can} \cdot t_\Delta^{*can}$. Then there is a formula $B \in t_\Delta^{*can}$ such that $B \to A \in s_\Delta^{*can}$. So $s : B \to A \in \Delta$ and $t : B \in \Delta$. Since Δ is a prime set, it is deductively closed, so it contains the axiom

$$s : (B \to A) \to (t : B \to s \cdot t : A).$$

Again since Δ is deductively closed, it follows by (MP) that $s \cdot t : A \in \Delta$, so $A \in (s \cdot t)^{-1}\Delta = (s \cdot t)_\Delta^{*can}$.

(2) $s_\Delta^{*can} \cup t_\Delta^{*can} \subseteq (s+t)_\Delta^{*can}$. Let $A \in s_\Delta^{*can} \cup t_\Delta^{*can}$. Case 1: $A \in s_\Delta^{*can} = s^{-1}\Delta$. Then $s : A \in \Delta$. Since Δ is deductively closed, it contains the axiom

$$s : A \to (s+t) : A.$$

Thus by (MP) we find $(s+t) : A \in \Delta$, i.e., $A \in (s+t)^{-1}\Delta = (s+t)_\Delta^{*can}$. The second case is analogous.

(3) $(t, A) \in \mathsf{CS} \implies A \in t_\Delta^{*can}$. By axiom necessitation we find that $\vdash t : A$, so $\Delta \vdash t : A$. Since Δ is deductively closed, it follows that $t : A \in \Delta$, so $A \in t^{-1}\Delta = t_\Delta^{*can}$.

(4) $s : s_\Delta^{*can} \subseteq (!s)_\Delta^{*can}$. Let $A \in s : s_\Delta^{*can}$. Then A is of the form $s : B$ for some formula $B \in s_\Delta^{*can} = s^{-1}\Delta$, i.e., $s : B \in \Delta$. We find that the axiom $(s : B) \to !s : (s : B) \in \Delta$, so $!s : (s : B) \in \Delta$, which means $s : B \in (!s)^{-1}\Delta = (!s)_\Delta^{*can}$.

Now we check the monotonicity conditions.

(M1) Assume that $p_\Gamma^{*can} = 1$ and $\Gamma \subseteq \Delta$. By the definition of $*^{can}$ we have that $p \in \Gamma$, so $p \in \Delta$ hence $p_\Delta^{*can} = 1$.

(M2) Now assume that $\Gamma \subseteq \Delta$. Then $t^{-1}\Gamma \subseteq t^{-1}\Delta$, which is $t_\Gamma^{*can} \subseteq t_\Delta^{*can}$. □

Lemma 4.13 (Truth Lemma). *For any formula A and any prime set Δ :*

$$A \in \Delta \iff (B^{can}, \Delta) \vDash A.$$

Proof. By induction on the formula A. We distinguish the following cases.

1. $A = p$ or $A = \bot$. By definition.

2. $A = B \wedge C$. Assume that $B \wedge C \in \Delta$. Since Δ is deductively closed, we have $B \in \Delta$ and $C \in \Delta$, so it follows by the induction hypothesis that $(B^{can}, \Delta) \vDash B$ and $(B^{can}, \Delta) \vDash C$.

   For the other direction assume that $(B^{can}, \Delta) \vDash B \wedge C$, so $(B^{can}, \Delta) \vDash B$ and $(B^{can}, \Delta) \vDash C$. By the induction hypothesis, we get that $B \in \Delta$ and $C \in \Delta$. Since Δ is deductively closed, it follows that $B \wedge C \in \Delta$.

3. $A = B \vee C$. Assume that $B \vee C \in \Delta$. Since Δ has the disjunction property, it follows that $B \in \Delta$ or $C \in \Delta$, so by the induction hypothesis, $(B^{can}, \Delta) \vDash B$ or $(B^{can}, \Delta) \vDash C$, so $(B^{can}, \Delta) \vDash B \vee C$.

For the other direction assume that $(B^{can}, \Delta) \vDash B \vee C$. Then

$$(B^{can}, \Delta) \vDash B \text{ or } (B^{can}, \Delta) \vDash C,$$

so by the induction hypothesis, $B \in \Delta$ or $C \in \Delta$. Since Δ is deductively closed, it follows that $B \vee C \in \Delta$.

4. $A = B \to C$. Assume that $B \to C \in \Delta$. We have to show $(B^{can}, \Delta) \vDash B \to C$, so let Γ be a prime set such that $\Delta \subseteq \Gamma$ and $(B^{can}, \Gamma) \vDash B$. It follows by the induction hypothesis that $B \in \Gamma$, and since $B \to C \in \Gamma$ and Γ is deductively closed, we have that $C \in \Gamma$. Applying the induction hypothesis again, we get that $(B^{can}, \Gamma) \vDash C$.

For the other direction assume that $(B^{can}, \Delta) \vDash B \to C$. We have to show that $B \to C \in \Delta$. Assume for a contradiction that $B \to C \notin \Delta$. Since Δ is deductively closed, it follows that $\Delta \nvdash B \to C$. It follows by the Deduction Theorem 3.6 that $\Delta \cup \{B\} \nvdash C$. By the Prime Lemma 4.9, there is a prime set Γ such that $\Delta \cup \{B\} \subseteq \Gamma$ and $\Gamma \nvdash C$, so in particular, $C \notin \Gamma$. By the induction hypothesis it follows that $(B^{can}, \Gamma) \vDash B$ and $(B^{can}, \Gamma) \nvDash C$, contradicting our assumption that $(B^{can}, \Delta) \vDash B \to C$.

5. $A = t : B$. We have

$$t : B \in \Delta \iff B \in t^{-1}\Delta = *^{can}(t, \Delta) \iff (B^{can}, \Delta) \vDash t : B.$$

$\square$

Lemma 4.14. B^{can} is a basic modular model.

Proof. We only have to show factivity, for which we use the Truth Lemma. Assume that

$$A \in *^{can}(t, \Delta) = t^{-1}\Delta.$$

By Lemma 4.10 we know that $t^{-1}\Delta \subseteq \Delta$, so we have $A \in \Delta$. By the Truth Lemma for the canonical basic modular model, we can conclude that $(B^{can}, \Delta) \vDash A$. So factivity is shown. $\square$

Theorem 4.15 (Completeness of iJT4$_{CS}$ with respect to basic modular models). For any formula A:

$$A \text{ is valid with respect to basic modular models } \quad \text{implies} \quad \vdash A.$$

Proof. By contraposition. Assume that $\not\vdash A$. By the Prime Lemma 4.9, there exists a prime set Δ such that $\Delta \not\vdash A$. In particular, $A \notin \Delta$. By the Truth Lemma 4.13, it follows that

$$(B^{can}, \Delta) \not\vDash A$$

since this structure is a basic modular model, it follows that A is not valid with respect to basic modular models. □

5 Modular Models

In this section, we introduce modular models for intuitionistic justification logic. Modular models are epistemic models in the sense that they feature possible worlds to model the notion of knowledge. The main principle of these logics is called *justification yields belief*, which means that if there is a justification for a formula A, then that formula must hold in all accessible worlds.

Modular models may seem too expressive as our language does not include a □-operator. However, these models explain the connection between implicit and explicit notions of belief. The main feature of modular models is that they provide a clear ontological separation of justification and truth, see, e.g., [6, 19].

In the second part of this section, we study so-called *fully explanatory* modular models. These models additionally require that if a formula holds in all accessible worlds, then there must be a justification for that formula. This principle can be seen as the reverse direction of justification yields belief.

Definition 5.1 (Quasimodels). A *quasimodel* is a tuple

$$\mathfrak{M} = (W, \leq, R, *),$$

such that $(W, \leq, *)$ is a basic evaluation, and R is a binary relation on W.

Definition 5.2 (Truth in quasimodels). We define what it means for a formula A to hold at a world $w \in W$ of a quasimodel $\mathfrak{M} = (W, \leq R, *)$, written $(\mathfrak{M}, w) \vDash A$, inductively as follows:

- $(\mathfrak{M}, w) \not\vDash \bot$;
- $(\mathfrak{M}, w) \vDash p$ iff $p_w^* = 1$;
- $(\mathfrak{M}, w) \vDash A \wedge B$ iff $(\mathfrak{M}, w) \vDash A$ and $(\mathfrak{M}, w) \vDash B$;
- $(\mathfrak{M}, w) \vDash A \vee B$ iff $(\mathfrak{M}, w) \vDash A$ or $(\mathfrak{M}, w) \vDash B$;

- $(\mathfrak{M}, w) \vDash A \to B$ iff $(\mathfrak{M}, v) \vDash B$ for all $v \geq w$ with $(\mathfrak{M}, v) \vDash A$;

- $(\mathfrak{M}, w) \vDash t : A$ iff $A \in t_w^*$.

Further we define $\Box_w := \{A \in \mathcal{L}_\mathsf{J} \mid (\mathfrak{M}, v) \vDash A \text{ for all } v \in R[w]\}$.

Lemma 5.3 (Locality of truth in quasimodels). Let $\mathfrak{B} = (W, \leq, *)$ be a basic evaluation and $\mathfrak{M} = (W, \leq, R, *)$ be a quasimodel. We find that for each $w \in W$ and each formula A,
$$(\mathfrak{M}, w) \vDash A \iff (\mathfrak{B}, w) \vDash A.$$

Definition 5.4 (Factive quasimodel). A quasimodel $\mathfrak{M} = (W, \leq, R, *)$ is called *factive* if $A \in t_w^*$ implies $(\mathfrak{M}, w) \vDash A$ for all $w \in W, t \in \mathsf{Tm}$, and formulas A.

Definition 5.5 (Modular models). A quasimodel $\mathfrak{M} = (W, \leq R, *)$ is called a *modular model* if it meets the following conditions:

(1) $t_w^* \subseteq \Box_w$ for all $t \in \mathsf{Tm}$ and $w \in W$ (JYB);

(2) R is reflexive;

(3) R is transitive;

(4) $w \leq v \implies R[v] \subseteq R[w]$ (Compatibility of $\leq$ with R).

We say that a formula A is *valid with respect to modular models* if for each modular model $\mathfrak{M} = (W, \leq, *)$ and all $w \in W$ we have $(\mathfrak{M}, w) \vDash A$.

The abbreviation JYB stands for *justification yields belief*, which is the main principle of modular models. This notion goes back to Artemov [6].

Lemma 5.6 (Modular models are factive). All modular models are factive.

Proof. Whenever $A \in t_w^*$ for some formula A, some $t \in \mathsf{Tm}$, and some $w \in W$, we have $A \in \Box_w$ by JYB. Since $R(w, w)$ by the reflexivity of R, we obtain $(\mathfrak{M}, w) \vDash A$ from the definition of $\Box_w$. □

Corollary 5.7 (Factivity of basic evaluations used in modular models). For any modular model $\mathfrak{M} = (W, \leq, R, *)$ we have that the basic evaluation $\mathfrak{B} := (W, \leq, *)$ is factive and, hence, a basic modular model.

Proof. Assume that for the basic evaluation $(W, \leq, *)$, we have $A \in t_w^*$ for some formula A, some point $w \in W$ and some term $t \in \mathsf{Tm}$. Then $A \in t_w^*$ in the modular model notation. By the previous lemma, we get $(\mathfrak{M}, w) \vDash A$, from which we conclude $(\mathfrak{B}, w) \vDash A$ by Lemma 5.3. □

Lemma 5.8 (Justifications remain relevant). Let $\mathfrak{M} = (W, \leq, R, *)$ be a modular model. Then for any $t \in \mathsf{Tm}$ and for arbitrary $w, v \in W$, if $R(w, v)$, then $t_w^* \subseteq t_v^*$, i.e., justifications remain relevant in accessible worlds.

Proof. Assume $R(w, v)$ and $A \in t_w^*$ for some formula A. Then we have $t : A \in (!t)_w^*$ because $(W, \leq, *)$ is a basic evaluation. Therefore, $t : A \in \square_w$ by JYB and, in particular, $(\mathfrak{M}, v) \vDash t : A$ by the definition of $\square_w$, which means that $A \in t_v^*$. □

Theorem 5.9 (Soundness and completeness: modular models). For any constant specification CS and any formula A we have

$$\vdash A \quad \Longleftrightarrow \quad A \text{ is valid with respect to modular models.}$$

Proof. Soundness. Let $\mathfrak{M} = (W, \leq, R, *)$ be a modular model. We need to show that any formula A such that $\vdash A$ holds at any world $w \in W$. By Corollary 5.7, we know that $\mathfrak{B} := (W, \leq, *)$ is a basic modular model. By soundness of iJT4$_\mathsf{CS}$ with respect to basic modular models, we get $(\mathfrak{B}, w) \vDash A$. Hence, $(\mathfrak{M}, w) \vDash A$ by the locality of truth in quasimodels (Lemma 5.3).

Completeness. For the opposite direction, suppose $\nvdash A$. By completeness of iJT4$_\mathsf{CS}$ with respect to basic modular models, there exists a basic modular model $\mathfrak{B} = (W, \leq, *)$ and a world $w \in W$ such that $(\mathfrak{B}, w) \nvDash A$. We define a quasimodel $\mathfrak{M} := (W, \leq, R, *)$ with $R := \leq$. By locality of truth for quasimodels (Lemma 5.3), we have that $(\mathfrak{M}, w) \nvDash A$, and it only remains to show that $\mathfrak{M}$ is a modular iJT4$_\mathsf{CS}$-model, i.e., that all the restrictions on R and the condition JYB are met. The reflexivity and transitivity of R are trivial. We check condition (4) (Compatibility of $\leq$ with R), i.e., $w \leq v \Longrightarrow R[v] \subseteq R[w]$. Assume $w \leq v$ and $u \in R[v]$. This means that $v \leq u$, so by transitivity of $\leq$ we have $w \leq u$ which means that $u \in R[w]$. Let us finish the proof by demonstrating JYB. Assume that $A \in t_w^*$ and $R(w, v)$. From this we get that $(\mathfrak{B}, w) \vDash t : A$ and $w \leq v$. By monotonicity for basic modular models, it follows that $(\mathfrak{B}, v) \vDash t : A$, so $A \in t_v^*$. By the factivity of basic modular models, we get that $(\mathfrak{B}, v) \vDash A$, and by the locality of truth in quasimodels, $(\mathfrak{M}, v) \vDash A$. Since v was arbitrary, we conclude that $A \in \square_w$. □

Definition 5.10 (Fully explanatory modular models). A modular model $\mathfrak{M} = (W, \leq, R, *)$ is *fully explanatory* if for any $w \in W$,

$$\square_w \subseteq \bigcup_{t \in \mathsf{Tm}} t_w^*,$$

i.e., $A \in \square_w$ implies $A \in t_w^*$ for some $t \in \mathsf{Tm}$.

We need the following auxiliary definition.

Definition 5.11. $\Gamma/\sharp := \{A \in \mathcal{L}_\mathsf{J} \mid t : A \in \Gamma \quad \text{for some } t \in \mathsf{Tm}\}$.

Lemma 5.12.
$$\Gamma \subseteq \Delta \implies \Gamma/\sharp \subseteq \Delta/\sharp$$

Proof. Assume that $\Gamma \subseteq \Delta$ and let $A \in \Gamma/\sharp$. By definition, there exists a term t, such that $t : A \in \Gamma$, so $t : A \in \Delta$ and $A \in \Delta/\sharp$. □

Lemma 5.13 (Soundness and completeness: fully explanatory modular models). Let CS be an axiomatically appropriate constant specification. Then iJT4$_\mathsf{CS}$ is sound and complete with respect to fully explanatory modular models.

Proof. Soundness immediately follows from soundness with respect to all modular models (and holds independently of whether CS is axiomatically appropriate).

We define the canonical model as
$$\mathfrak{M}^{can} := (W^{can}, \leq^{can}, R^{can}, *^{can})$$

where

(i) $W^{can} := \{\Delta \subseteq \mathcal{L}_\mathsf{J} \mid \Delta \text{ is prime}\}$

(ii) $\leq^{can} := \subseteq$

(iii) $*^{can}(p, \Delta) = 1$ iff $p \in \Delta$

(iv) $*^{can}(t, \Delta) := t^{-1}\Delta$

(v) $R^{can}(\Gamma, \Delta)$ iff $\Gamma/\sharp \subseteq \Delta$

To show that $\mathfrak{M}^{can}$ is a modular iJT4$_\mathsf{CS}$-model, it remains to establish that the set W^{can} is non-empty, that R^{can} is reflexive and transitive, that $\leq^{can}$ is compatible with R^{can} and that the condition JYB is satisfied. We start with showing $W^{can} \neq \varnothing$. We have already shown that the empty set is iJT4$_\mathsf{CS}$-consistent, so by the Prime Lemma 4.9, there exists a prime set extending $\varnothing$, which is an element of W^{can}.

To show that $\leq^{can}$ is compatible with R^{can}, assume that $\Gamma \subseteq \Delta$. We need to show that $R^{can}[\Delta] \subseteq R^{can}[\Gamma]$, so we pick $\Pi \in R^{can}[\Delta]$ and show that $\Pi \in R^{can}[\Gamma]$. $\Pi \in R^{can}[\Delta]$ means that $\Delta/\sharp \subseteq \Pi$. By the lemma above, we have that $\Gamma/\sharp \subseteq \Delta/\sharp$, and therefore $\Gamma/\sharp \subseteq \Pi$, i.e., $\Pi \in R^{can}[\Gamma]$.

To show JYB, assume $A \in t^{*^{can}}_\Gamma$ for some formula A, some $t \in \mathsf{Tm}$, and some $\Gamma \in W^{can}$. We need to show that $A \in \Box_\Gamma$, i.e., that $(\mathfrak{M}^{can}, \Delta) \vDash A$ whenever $R^{can}(\Gamma, \Delta)$. Consider any such $\Delta \in W^{can}$. We have $t : A \in \Gamma$ by the definition of $t^{*^{can}}_\Gamma$ and $A \in \Delta$ by the definition of R^{can}. By the truth lemma for basic evaluations, it follows that $(\mathfrak{B}^{can}, \Delta) \models A$ where $\mathfrak{B}^{can} = (W^{can}, \leq^{can}, *^{can})$. By the locality of truth in quasimodels, we have $(\mathfrak{M}^{can}, \Delta) \models A$.

To show that R^{can} is reflexive, consider any $\Gamma \in W^{can}$. Assume that $A \in \Gamma/\sharp$, i.e., that $t : A \in \Gamma$ for some $t \in \mathsf{Tm}$. Since Γ is prime, it is deductively closed. $t : A \to A$ is an axiom, so $t : A \to A \in \Gamma$. Again, since Γ is deductively closed, it follows by (MP) that $A \in \Gamma$. Therefore, $\Gamma/\sharp \subseteq \Gamma$, which means that $R^{can}(\Gamma, \Gamma)$.

To show that R^{can} is transitive, consider arbitrary $\Gamma, \Delta, \Pi \in W^{can}$ such that $R^{can}(\Gamma, \Delta)$ and $R^{can}(\Delta, \Pi)$. Assume that $A \in \Gamma/\sharp$, i.e., that $t : A \in \Gamma$ for some $t \in \mathsf{Tm}$. Since Γ is prime, it is deductively closed, and since $t : A \to !t : t : A$ is an axiom of iJT4$_{\mathsf{CS}}$, we conclude $!t : t : A \in \Gamma$. Hence $t : A \in \Gamma/\sharp \subseteq \Delta$ and $A \in \Delta/\sharp \subseteq \Pi$. Therefore, $\Gamma/\sharp \subseteq \Pi$, which means $R^{can}(\Gamma, \Pi)$.

Finally, we show that $\mathfrak{M}^{can}$ is fully explanatory. Assume that $A \in \Box_\Gamma$ for some formula A and prime set Γ. Then

$$\Gamma/\sharp \vdash A \tag{1}$$

Indeed, assume for a contradiction that $\Gamma/\sharp \nvdash A$. By the Prime Lemma, there exists a prime set Δ such that $\Gamma/\sharp \subseteq \Delta$ and $\Delta \nvdash A$. By the definition of R^{can}, we have $R^{can}(\Gamma, \Delta)$, and from $\Delta \nvdash A$ we get that $A \notin \Delta$. By the Truth Lemma for basic evaluations, it follows that $(\mathfrak{B}^{can}, \Delta) \nvDash A$. By the locality of truth in quasimodels, we have $(\mathfrak{M}^{can}, \Delta) \nvDash A$, contradicting our assumption that $A \in \Box_\Gamma$. By (1), it follows that there are a finite set $G_1, \ldots, G_n \in \Gamma/\sharp$, such that

$$G_1, \ldots, G_n \vdash A.$$

Since each $G_i \in \Gamma/\sharp$, there must exist terms $s_i \in \mathsf{Tm}$ such that $s_i : G_i \in \Gamma$ for each $1 \leq i \leq n$.

By Lemma 3.7, given the axiomatic appropriateness of CS, there exists a term t such that

$$s_1 : G_1, \ldots, s_n : G_n \vdash t : A$$

By the Deduction Theorem

$$\vdash s_1 : G_1 \to (s_2 : G_2 \to \cdots \to (s_n : G_n \to t : A) \ldots).$$

Γ is prime, so it is deductively closed, and therefore $t : A \in \Gamma$ and finally

$$A \in t^{-1}\Gamma = t_\Gamma^{*can}.$$

So $\mathfrak{M}^{can}$ is fully explanatory. $\square$

6 Realization

We establish in this section that the justification logic iJT4 is the explicit counterpart of the intuitionistic modal logic iS4. This is simply a reformulation of [3, Section 3] using axiomatically appropriate and schematic constant specifications.

First we show that iS4 is the forgetful projection of iJT4. We need the following definition: if A is a formula of $\mathcal{L}_J$, then A° is the formula of $\mathcal{L}_I$ that is the result of replacing all occurrences of $t:$ in A with $\Box$. We immediately get the following theorem.

Theorem 6.1 (Forgetful projection). Let CS be an arbitrary constant specification. For each $\mathcal{L}_J$-formula A,

$$\vdash_{\mathsf{iJT4}_{\mathsf{CS}}} A \quad \text{implies} \quad \vdash_{\mathsf{HiS4}} A^\circ.$$

Proof. By induction on the length of the $\mathsf{iJT4}_{\mathsf{CS}}$ derivation.

It is easy to see that for each axiom A of $\mathsf{iJT4}_{\mathsf{CS}}$, we have $\vdash_{\mathsf{HiS4}} A^\circ$.

If A is the conclusion of an application of modus ponens from premises B and $B \to A$, then by induction hypothesis and the definition of $\cdot^\circ$ we get

$$\vdash_{\mathsf{HiS4}} B^\circ \quad \text{and} \quad \vdash_{\mathsf{HiS4}} B^\circ \to A^\circ$$

and thus $\vdash_{\mathsf{HiS4}} A^\circ$ by modus ponens.

If A is the conclusion of an instance of axiom necessitation, then A has the form $c : B$ for some axiom B of $\mathsf{iJT4}_{\mathsf{CS}}$. Therefore, as shown above, $\vdash_{\mathsf{HiS4}} B^\circ$. An application of necessitation yields $\vdash_{\mathsf{HiS4}} \Box B^\circ$, which is $\vdash_{\mathsf{HiS4}} A^\circ$. □

Now we show the converse direction, namely that iJT4 realizes iS4. For this, we need the following definition: a *realization* r is a mapping from $\mathcal{L}_I$ to $\mathcal{L}_J$ such that for each $\mathcal{L}_I$-formula A we have that

$$(r(A))^\circ = A.$$

A realization is *normal* if all negative occurrences of $\Box$ are realized by justification variables.

Theorem 6.2 (Realization). Let CS be an axiomatically appropriate and schematic constant specification. Then there exists a realization r such that for each $\mathcal{L}_I$-formula A we have

$$\vdash_{\mathsf{GiS4}} A \quad \text{implies} \quad \vdash_{\mathsf{iJT4}_{\mathsf{CS}}} r(A).$$

Proof. It turns out that Artemov's original realization proof for LP [2] also works in the intuitionistic case. We will only give a proof sketch here.

We start with defining positive and negative occurrences of $\Box$ in a sequent as usual. Observe that the rules of GiS4 respect these polarities so that $(\supset \Box)$ introduces positive occurrences of $\Box$ and $(\Box \supset)$ introduces negative occurrences of $\Box$. Occurrences of $\Box$ are *related* if they occur in related formulas of premises and conclusions of rules; we close this relationship of related occurrences under transitivity. All occurrences of $\Box$ in a GiS4-derivation naturally split into disjoint *families* of related occurrences. We call a family *essential* if at least one of its members is introduced by a $(\supset \Box)$ rule. Note that an essential family is positive (i.e. contains only positive occurrences).

Now let $\mathcal{D}$ be the GiS4 derivation that proves A. The desired $\mathcal{L}_J$-formula $r(A)$ is constructed by the following three steps. We reserve a large enough set of justification variables as *provisional variables*.

1. For each negative family and each non-essential positive family, replace all $\Box$ occurrences by x : where we choose a fresh justification variable for each family.

2. Pick an essential family f. Enumerate all occurrences of $(\supset \Box)$ rules that introduce a $\Box$-operator to this family. Replace each $\Box$ with a justification term
$$v_1 + \cdots + v_{n_f}$$
where each v_i is a fresh provisional variable. Do this for each essential family. The resulting tree $\mathcal{D}'$ is labelled by $\mathcal{L}_J$-formulas.

3. Replace the provisional variables starting with the leaves and working toward the root. By induction on the depth of a node in $\mathcal{D}'$ we establish that after the process passes a node, the sequent assigned to this node becomes derivable in iJT4$_{CS}$ where derivability of $\Gamma \supset A$ means $\Gamma \vdash_{\mathsf{iJT4}_{CS}} A$. We distinguish the following cases.

    (a) The axioms $\Gamma \supset A$ with $A \in \Gamma$ or $\bot \in \Gamma$ are derivable in iJT4$_{CS}$.

    (b) For every rule other than $(\supset \Box)$ we do not change the term assignment and establish that the conclusion of the rule is derivable in iJT4$_{CS}$ if the premises are.

    (c) Let an occurrence of a $(\supset \Box)$ rule have number i in the enumeration of all $(\supset \Box)$ rules in a given family f. The corresponding node in $\mathcal{D}'$ is labelled

by
$$\frac{y_1 : B_1, \ldots, y_k : B_k \supset A}{y_1 : B_1, \ldots, y_k : B_k \supset u_1 + \cdots + u_{n_f} : A}$$

where the y's are justification variables, the u's are justification terms, and u_i is a provisional variable. By the induction hypothesis

$$y_1 : B_1, \ldots, y_k : B_k \supset A$$

is derivable in iJT4$_{CS}$. Using the Lifting Lemma, we construct a term t such that

$$y_1 : B_1, \ldots, y_k : B_k \vdash_{\mathsf{iJT4_{CS}}} t : A.$$

Thus

$$y_1 : B_1, \ldots, y_k : B_k \vdash_{\mathsf{iJT4_{CS}}} u_1 + \cdots + u_{i-1} + t + u_{i+1} + \cdots + u_{n_f} : A.$$

Substitute t for u_i everywhere in $\mathcal{D}'$. By Lemma 3.9, this does not affect the already established derivability results.

Eventually, all provisional variables are replaced with terms of non-provisional variables in $\mathcal{D}'$ and we have established that its root sequent $r(A)$ is derivable in iJT4$_{CS}$. The realization r built by this construction is normal. $\square$

7 Conclusion

We have established that if we take the classical Logic of Proofs and change the underlying classical propositional logic to intuitionistic propositional logic, then we obtain an explicit counterpart of the intuitionistic modal logic iS4. This is an interesting result since the logic of proofs of Heyting arithmetic includes additional axioms that introduce special justification terms for all admissible rules of intuitionistic logic. This seems necessary to obtain completeness with respect to provability semantics where the justification relation is interpreted by formal provability in Heyting Arithmetic.

Our results now show that these additional axioms and justification terms are not needed if we are interested in the explicit counterpart of intuitionistic modal logic and the corresponding possible world semantics for justification logic.

Moreover, we believe that intuitionistic justification logics will help to understand intuitionistic modal logics better. In particular, they will help to clarify the role of additional principles for the $\square$-modality and the corresponding conditions on the accessibility relation. However, this is left for future research.

References

[1] Sergei N. Artemov. Operational modal logic. Technical Report MSI 95–29, Cornell University, December 1995.

[2] Sergei N. Artemov. Explicit provability and constructive semantics. *Bulletin of Symbolic Logic*, 7(1):1–36, March 2001.

[3] Sergei N. Artemov. Unified semantics for modality and λ-terms via proof polynomials. In Kees Vermeulen and Ann Copestake, editors, *Algebras, Diagrams and Decisions in Language, Logic and Computation*, volume 144 of *CSLI Lecture Notes*, pages 89–118. CSLI Publications, Stanford, 2002.

[4] Sergei [N.] Artemov. Justified common knowledge. *Theoretical Computer Science*, 357(1–3):4–22, July 2006.

[5] Sergei [N.] Artemov. The logic of justification. *The Review of Symbolic Logic*, 1(4):477–513, December 2008.

[6] Sergei N. Artemov. The ontology of justifications in the logical setting. *Studia Logica*, 100(1–2):17–30, April 2012. Published online February 2012.

[7] Sergei [N.] Artemov and Rosalie Iemhoff. The basic intuitionistic logic of proofs. *Journal of Symbolic Logic*, 72(2):439–451, June 2007.

[8] Sergei [N.] Artemov and Elena Nogina. Introducing justification into epistemic logic. *Journal of Logic and Computation*, 15(6):1059–1073, December 2005.

[9] Alexandru Baltag, Bryan Renne, and Sonja Smets. The logic of justified belief, explicit knowledge, and conclusive evidence. *Annals of Pure and Applied Logic*, 165(1):49–81, January 2014. Published online in August 2013.

[10] Samuel Bucheli, Roman Kuznets, and Thomas Studer. Justifications for common knowledge. *Journal of Applied Non-Classical Logics*, 21(1):35–60, January–March 2011.

[11] Samuel Bucheli, Roman Kuznets, and Thomas Studer. Partial realization in dynamic justification logic. In Lev D. Beklemishev and Ruy de Queiroz, editors, *Logic, Language, Information and Computation, 18th International Workshop, WoLLIC 2011, Philadelphia, PA, USA, May 18–20, 2011, Proceedings*, volume 6642 of *Lecture Notes in Artificial Intelligence*, pages 35–51. Springer, 2011.

[12] Samuel Bucheli, Roman Kuznets, and Thomas Studer. Realizing public announcements by justifications. *Journal of Computer and System Sciences*, 80(6):1046–1066, 2014.

[13] Evgenij Dashkov. Arithmetical completeness of the intuitionistic logic of proofs. *Journal of Logic and Computation*, 21(4):665–682, August 2011. Published online August 2009.

[14] G. Fischer Servi. Axiomatizations for some intuitionistic modal logics. *Rendiconti del Seminario Matematico Universita' e Politecnico di Torino*, 42:179–194, 1984.

[15] Melvin Fitting. The logic of proofs, semantically. *Annals of Pure and Applied Logic*, 132(1):1–25, February 2005.

[16] Meghdad Ghari. Pavelka-style fuzzy justification logics. *Logic Journal of the IGPL*, 24(5):743–773, 2016.

[17] Gerhard Jäger and Michel Marti. Intuitionistic common knowledge or belief. *Journal*

of Applied Logic, 18:150–163, 2016.

[18] Ioannis Kokkinis, Petar Maksimović, Zoran Ognjanović, and Thomas Studer. First steps towards probabilistic justification logic. *Logic Journal of IGPL*, 23(4):662–687, 2015.

[19] Roman Kuznets and Thomas Studer. Justifications, ontology, and conservativity. In Thomas Bolander, Torben Braüner, Silvio Ghilardi, and Lawrence Moss, editors, *Advances in Modal Logic, Volume 9*, pages 437–458. College Publications, 2012.

[20] Roman Kuznets and Thomas Studer. Update as evidence: Belief expansion. In Sergei [N.] Artemov and Anil Nerode, editors, *Logical Foundations of Computer Science, International Symposium, LFCS 2013, San Diego, CA, USA, January 6–8, 2013, Proceedings*, volume 7734 of *Lecture Notes in Computer Science*, pages 266–279. Springer, 2013.

[21] Roman Kuznets and Thomas Studer. Weak arithmetical interpretations for the logic of proofs. *Logic Journal of the IGPL*, 24(3):424–440, 2016.

[22] Hiroakira Ono. On some intuitionistic modal logics. *Publications of the Research Institute for Mathematical Sciences*, 13(3):687–722, 1977.

[23] Gordon Plotkin and Colin Stirling. A framework for intuitionistic modal logics: Extended abstract. In *Proceedings of the 1986 Conference on Theoretical Aspects of Reasoning About Knowledge*, TARK '86, pages 399–406. Morgan Kaufmann Publishers Inc., 1986.

[24] Alex K. Simpson. *The proof theory and semantics of intuitionistic modal logic*. PhD thesis, University of Edinburgh, 1994.

The Indian Schema Analogy Principles

J.B.Paris*
School of Mathematics, The University of Manchester, Manchester M13 9PL.
jeff.paris@manchester.ac.uk

A.Vencovská[†]
School of Mathematics, The University of Manchester, Manchester M13 9PL.
alena.vencovska@manchester.ac.uk

Abstract

We investigate the status within Unary Pure Inductive Logic of a family of analogy principles suggested by the so called Indian Schema from Gotama's Nyāyasūtra showing that they all follow from the symmetry principle of Atom Exchangeability. Their status under the weaker assumptions of Constant and Predicate Exchangeability and Strong Negation are also investigated.

Keywords: Indian Schema, Nyāyasūtra, Analogy, Pure Inductive Logic, Logical Probability, Rationality, Uncertain Reasoning.

Introduction

In the Nyāyasūtra of Gotama[1], the founding Indian logic text from c.200BCE-150CE, the author aims to delineate in five terse aphorisms (Sūtras 3.1.32,3.1.34-37) a scheme for right reasoning. Subsequently numerous commentators, most notably Vātsyāyana (c.375CE-450CE), Uddyotakara (6th century CE), Gaṅgeśa (12th century CE), added their own explanations and developments, as well as incorporating revisions from other Indian Schools of Philosophy. When H.T.Colebrooke first introduced this Nyāya system of philosophy to the Victorian public at a meeting of the Royal Asiatic Society in London in 1824 the pattern of reasoning he called the

---

We would like to thank the referees for their comments, especially for correcting us on several points related to the history and language of the Nyāyasūtra.

*Supported by a UK Engineering and Physical Sciences Research Council Research Grant.
[†]Supported by a UK Engineering and Physical Sciences Research Council Research Grant.
[1]Aka Gautama, Akṣapāda.

Hindu Syllogism, subsequently dubbed the Indian Schema, was exemplified by a handful of examples,[2] most prominently the Smoke-Fire example:[3]

(a) *Where there is smoke there is fire, like in the kitchen.*

(b) *There is smoke on the hill.*

(c) *Therefore there is fire on the hill.*

How exactly such examples should be understood, in their classical Sanskrit context or within the framework of contemporary notions of reasoning (e.g. deductive, default, case based, etc.) has been the subject of much debate, see [3] for a sample. In particular it is a moot point why the instance 'like in the kitchen' is present at all since the earlier 'Where there is smoke there is fire' would seem to render it redundant. Certainly its featuring there was taken by some Victorians to dismiss the Indian Schema as simply invalid analogical reasoning, from particular to particular.

In a previous paper, [11], we mentioned possible grounds for supposing that Gotama's original intention for line (a) (the udāharaṇa) was just to cite an instance of smoke in a kitchen being the result of a fire and that the later explicit introduction of the universal by the Buddhist logician Dharmakīrti, see Oetke's [10], or possibly his predecessor Dignāga (c.480CE-540CE), see Ganeri's [5, page 38],[4] represented a shift from analogical to deductive reasoning. Viewed in this way then the Indian Schema becomes:

(a) *When there was smoke in the kitchen there was fire.*

(b) *There is smoke on the hill.*

(c) *Therefore there is fire on the hill.* $\hspace{2cm}(\star)$

Whether or not this was Gotama's original intention, we continued in [11] to argue that nevertheless, and not withstanding the Victorians' rebuke, the 'reasoning' in this version can be justified as *rational* within the context of Pure Inductive Logic (hereafter PIL) – see [12]. Our interest in this paper is to investigate more fully some of the analogical principles of probability assignment resulting from various possible formalisations of $(\star)$. Before doing so we need to briefly recall the framework of PIL as presented, for example, in [12].

---

[2] For a quick introduction see J.Ganeri's [4].

[3] This is the (last) three line form meant for reasoning for oneself, see [13].

[4] Ganeri also points out in this paper that Vātsyāyana (c.375CE-450CE) had already made a significant step in this direction.

The Pure Inductive Logic Context

Pure Inductive Logic as described in [12] is conventionally set within a predicate language L with a finite set of relation symbols and countably many constant symbols $a_1, a_2, a_3, \ldots$ and no function symbols nor equality. Let SL denote the set of sentences of L_q and let $QFSL$ denote the quantifier free sentences in L.

A probability function on L is a function $w : SL \to [0,1]$ such that for $\theta, \phi, \exists x\, \psi(x) \in SL$,

(i) If $\models \theta$ then $w(\theta) = 1$.

(ii) If $\theta \models \neg \phi$ then $w(\theta \vee \phi) = w(\theta) + w(\phi)$.

(iii) $w(\exists x\, \psi(x)) = \lim_{n \to \infty} w\left(\bigvee_{i=1}^{n} \psi(a_i)\right)$.

From these all the expected properties of probability follow (see [12, Proposition 3.1]), in particular if $\theta \models \phi$ then $w(\theta) \leq w(\phi)$.

Given such a w we set the conditional probability function $w(\cdot \mid \cdot) : SL \times SL \to [0,1]$ to be a function such that for $\theta, \phi \in SL$ with $w(\phi) > 0$,

$$w(\theta \mid \phi) = \frac{w(\theta \wedge \phi)}{w(\phi)}.$$

In practice it will be convenient to identify

$$w(\theta \mid \phi) = c \quad \text{with} \quad w(\theta \wedge \phi) = c w(\phi)$$

since this avoids separating the cases when $w(\phi)$ is zero and non-zero.

In PIL we are, at this stage of its development, primarily interested in elucidating 'rationality constraints' on w in the case when the symbols of L are entirely uninterpreted. So if w is to represent a 'rational' assignment of probabilities to the sentences of L in the absence of any particular meaning of, and information about, the constants and the predicates what properties in addition to (i)-(iii) should w satisfy?

Numerous such constraints, usually in the form of principles that w should arguably obey, have been proposed based on various intuitions of what 'rational' might mean but the most forceful, going back to Johnson [7] and Carnap [1] (or see Carnap's Axioms for Inductive Logic at [14, p973]), are those based on symmetry, the idea being that it would be irrational of w to break existing symmetries in the language. At its simplest level this has been understood as saying that if we have an isomorphism of the symbols of a language then the probability assigned to a sentence should be the same as that assigned to its symbol-wise image under that isomorphism, because the isomorphism provides, or witnesses, a symmetry between sentences and their images.

The most obvious example of such a symmetry is when the isomorphism simply permutes the constant symbols and leaves the relation symbols fixed. In this case the requirement of preserving symmetries, that is of assigning the same probability to a sentence as to its isomorphic image, amounts to[5] :

The Constant Exchangeability Principle, Ex

For $\theta \in SL$ and constant symbols a_i, a_j of L, $w(\theta) = w(\theta')$ where θ' is the result of transposing a_i and a_j in θ.

Analogous to Constant Exchangeability but this time permuting relation symbols of the same arity gives:

The Predicate Exchangeability Principle, Px

For $\theta \in SL$ and relation symbols P_i, P_j of L of the same arity, $w(\theta) = w(\theta')$ where θ' is the result of transposing P_i and P_j in θ.

Satisfying these two principles is frequently viewed as necessary for w to be considered rational. A third symmetry condition is based on the idea that since the context is supposed to be entirely uninterpreted there is symmetry between R and $\neg R$,[6] just in the same way as there is between heads and tails when we toss a coin. This yields:

The Strong Negation Principle, SN

$w(\theta) = w(\theta')$ where θ' is the result of replacing each occurrence of the relation symbol R in θ by $\neg R$.

Until fairly recently Inductive Logic has been almost entirely concerned with unary languages, that is where all the relation (or predicate) symbols have a single argument. In this case there is a further widely accepted[7] symmetry principle, Atom Exchangeability. To wit let L_q be the language whose only relation (i.e. predicate) symbols are the unary $P_1, P_2, \ldots, P_q$ and let $\alpha_1(x), \alpha_2(x), \ldots, \alpha_{2^q}(x)$ be the *atoms*

---

[5]We remark that each of the constant symbols a_i, a_j in the formulation of Ex may but does not need to, feature in θ, and similarly for Px.

[6]Since $\neg\neg R \equiv R$. Again there is an underlying, albeit more complicated, isomorphism.

[7]Since it holds for the members of Carnap's Continuum of Inductive methods

of L_q, that is the formulae of the form

$$P_1^{\epsilon_1}(x) \wedge P_2^{\epsilon_2}(x) \wedge \ldots \wedge P_q^{\epsilon_q}(x)$$

where $\epsilon_1, \epsilon_2, \ldots, \epsilon_q \in \{0, 1\}$ and $P_i^\epsilon = P_i$ if $\epsilon = 1$, $\neg P_i$ if $\epsilon = 0$.

The Atom Exchangeability Principle, Ax

If σ is a permutation of $\{1, 2, \ldots, 2^q\}$ then

$$w\left(\bigwedge_{i=1}^n \alpha_{h_i}(a_i)\right) = w\left(\bigwedge_{i=1}^n \alpha_{\sigma(h_i)}(a_i)\right) \tag{1}$$

In the case of Ax then the 'symmetry' is between possible complete descriptions of constants – knowing which atom a particular a_i satisfies tells us everything there is to know about a_i as such. [For more on the purported 'rationality' of this principle see [12, p87].] Notice that both Px and SN follow from Ax.

In the case of this purely unary language L_q Constant Exchangeability, Ex, has two consequences which we shall be needing later. The first is de Finetti's Representation Theorem (in the context of this paper). To explain this let

$$\mathbb{D}_{2^q} = \{\langle x_1, x_2, \ldots, x_{2^q} \rangle \mid x_i \geq 0, \sum_i x_i = 1\}$$

and for $\vec{x} \in \mathbb{D}_{2^q}$ let

$$w_{\vec{x}}\left(\bigwedge_{i=1}^n \alpha_{h_i}(a_i)\right) = \prod_{i=1}^n x_{h_i}.$$

As shown in [12] $w_{\vec{x}}$ extends to a probability function on L_q which satisfies Ex. Indeed every probability function on L_q satisfying Ex is a convex combination of these $w_{\vec{x}}$:

de Finetti's Representation Theorem

If w is a probability function on L_q satisfying Ex then there is a countably additive normalized measure μ on $\mathbb{D}_{2^q}$ such that for any $\theta(a_1, a_2, \ldots, a_n) \in SL_q$,

$$w(\theta(a_1, a_2, \ldots, a_n)) = \int_{\mathbb{D}_{2^q}} w_{\vec{x}}(\theta(a_1, a_2, \ldots, a_n))\, d\mu,$$

and conversely.

We refer to μ as the de Finetti prior of w. If w additionally satisfies Ax then it may be assumed that the measure μ is invariant under permutations of the coordinates, see [12, Chapter 14].

A consequence of this theorem, due to Gaifman, [2] or see [12, p71], which we shall also need later is:

The Extended Principle of Instantial Relevance, EPIR

For $\theta(a_1, a_2, \ldots, a_n), \phi(a_1) \in SL_q$,

$$w(\phi(a_{n+2}) \,|\, \phi(a_{n+1}) \wedge \theta(a_1, a_2, \ldots, a_n)) \geq w(\phi(a_{n+2}) \,|\, \theta(a_1, a_2, \ldots, a_n)). \quad (2)$$

Two particular probability functions, $c_\infty^{L_q}$ and $c_0^{L_q}$, will figure in some of the results which follow.

Carnap's $c_\infty^{L_q}$ (aka m^*) equals $w_{\vec{x}}$ when

$$\vec{x} = \langle 2^{-q}, 2^{-q}, \ldots, 2^{-q} \rangle \in \mathbb{D}_{2^q}.$$

This probability function has the property that for any sentence $\theta(a_1) \in SL_q$ and $n \geq 0$,

$$c_\infty^{L_q}(\theta(a_{n+1}) \,|\, \theta(a_1) \wedge \theta(a_2) \wedge \ldots \wedge \theta(a_n)) = c_\infty^{L_q}(\theta(a_{n+1})). \quad (3)$$

That is, $c_\infty^{L_q}$ denies any inductive support, the probability it gives to $\theta(a_{n+1})$ is unaffected by the evidence that $\theta(a_i)$ held for $i = 1, 2, \ldots, n$ no matter how large n may be.

Carnap's $c_0^{L_q}$ equals

$$2^{-q} \sum_{i=1}^{2^q} w_{\vec{e_i}}$$

where $\vec{e_i} \in \mathbb{D}_{2^q}$ is the vector with 1 in the ith coordinate and 0's elsewhere. This probability function has the property that for any sentence $\theta(a_1)$ and $n > 0$,

$$c_0^{L_q}(\theta(a_{n+1}) \,|\, \theta(a_1) \wedge \theta(a_2) \wedge \ldots \wedge \theta(a_n)) = c_0^{L_q}(\theta(a_{n+1}) \,|\, \theta(a_1)) = 1.^8 \quad (4)$$

In this case then $c_0^{L_q}$ derives the maximal possible inductive support for $\theta(a_{n+1})$ on the basis of a single given $\theta(a_1)$. So as the notation already hints it is at the other end of the 'inductive scale' from $c_\infty^{L_q}$.

On account of the above properties neither $c_\infty^{L_q}$ nor $c_0^{L_q}$ are particularly favoured choices for blank slate probability assignments in the absence of any intended interpretation.

---

[8] Recall the convention regarding zero divisors in conditional probabilities.

Unary Formalisations

To make the forthcoming results more immediate we will write S, F, h, k for P_1, P_2, a_1, a_2.[9]

As argued in [11] for w a probability function on L_2[10] we might formalise $(\star)$ in one of the forms[11],[12]

$$w(F(h) \mid (S(k) \to F(k)) \wedge S(h)) > 1/2, \tag{5}$$

$$w(F(h) \mid (S(k) \leftrightarrow F(k)) \wedge S(h)) > 1/2, \tag{6}$$

$$w(F(h) \mid S(k) \wedge F(k) \wedge S(h)) > 1/2, \tag{7}$$

since a probability of more than $1/2$ could, in the uninterpreted, ceteris paribus, context of PIL, be taken as a justification for opting for $F(h)$ in the sense that it must be more probable than $\neg F(h)$. In [11] we showed that with the exception of $w = c_\infty^{L_2}$, any w satisfying Ex+Px+SN must also satisfy each of (5), (6) and (7); for $w = c_\infty^{L_2}$ the inequality in each of (5), (6) and (7) becomes equality.[13] So excluding $w = c_\infty^{L_2}$, each of (5), (6) and (7) is at least as rational as Ex+Px+SN.

Hence the only way one could argue that while accepting Ex+Px+SN, $(\star)$ as formalised by (5), (6) or (7) was not justified in PIL, would be if one held that $c_\infty^{L_2}$ was an acceptable choice. But then since, from (3),

$$c_\infty^{L_2}(F(a_{n+1}) \mid F(a_1) \wedge F(a_2) \wedge \ldots, \wedge F(a_n)) = c_\infty^{L_2}(F(a_{n+1})) = 1/2$$

one would also have to argue that even under the evidence of $F(a_1), F(a_2), \ldots, F(a_n)$ the conclusion that $F(a_{n+1})$ was not justified, no matter how large n was.

Buoyed by the apparent success that (5), (6) and (7) follow from Ex+Px+SN (unless $w = c_\infty^{L_2}$) one might naturally raise the question whether the belief that there is fire on the hill should not be greater the more kitchen fires one has experienced.

---

[9] Since we will always have Ex+Px the choices of suffices here are not important.

[10] We could equally take L_q ($q \geq 2$) here in place of L_2.

[11] We shall consider other possibilities later.

[12] In [11] we employed $\geq$ rather than $>$ but the strict inequality obviously carries more weight. Since the equality occurs only in very special cases, as discussed below, we prefer to adopt the strict inequality here.

[13] If instead we had taken L_q in place of L_2 the exceptions would have been those w which equal $c_\infty^{L_2}$ when restricted to SL_2.

In other words should not (5), (6) and (7) for w a probability function on L_2 be enhanced to

$$w(F(h) \mid \bigwedge_{i=1}^{n+1} (S(k_i) \to F(k_i)) \wedge S(h)) > w(F(h) \mid \bigwedge_{i=1}^{n} (S(k_i) \to F(k_i)) \wedge S(h)), \quad (8)$$

$$w(F(h) \mid \bigwedge_{i=1}^{n+1} (S(k_i) \leftrightarrow F(k_i)) \wedge S(h)) > w(F(h) \mid \bigwedge_{i=1}^{n} (S(k_i) \leftrightarrow F(k_i)) \wedge S(h)), \quad (9)$$

$$w(F(h) \mid \bigwedge_{i=1}^{n+1} (S(k_i) \wedge F(k_i)) \wedge S(h)) > w(F(h) \mid \bigwedge_{i=1}^{n} (S(k_i) \wedge F(k_i)) \wedge S(h)), \quad (10)$$

for $n \geq 0$? (Under the assumption of Ex+Px+SN, these are indeed enhancements since these principles imply $w(F(h) \mid S(h)) = 1/2$.)

Our plan now is to relate these particular 'Indian Schema Principles' (8), (9) and (10) to the established symmetry principles stated in the previous section. We start with Atom Exchangeability Ax.

Assuming Atom Exchangeability

It turns out that (8), (9) and (10) all essentially follow from Ex+Ax. Indeed the following stronger result holds:

Theorem 1. *For w a probability function on L_q satisfying Ex+Ax, $\theta(a_1), \phi(a_1) \in QFSL_q$,*

$$w(\theta(a_{n+2}) \mid \phi(a_{n+2}) \wedge \bigwedge_{i=1}^{n+1} \theta(a_i)) \geq w(\theta(a_{n+1}) \mid \phi(a_{n+1}) \wedge \bigwedge_{i=1}^{n} \theta(a_i)),$$

with equality just if $\theta(a_i) \wedge \phi(a_i)$ is inconsistent or $\neg \theta(a_i) \wedge \phi(a_i)$ is inconsistent or $w = c_\infty^{L_q}$ or $w = c_0^{L_q}$ and $n > 0$.

Proof. It is straightforward to check that if $\theta(a_i) \wedge \phi(a_i)$ is inconsistent or $\neg \theta(a_i) \wedge \phi(a_i)$ is inconsistent then the result holds with equality so assume that neither of these hold. Let the de Finetti representation of w satisfying Ax be

$$w = \int w_{\vec{x}} \, d\mu$$

where μ is invariant under permutations of the coordinates.

Without loss of generality let

$$\theta(a_1) \equiv \bigvee_{i=1}^{r} \alpha_i(a_1), \quad \phi(a_1) \equiv \bigvee_{i=1}^{m} \alpha_i(a_1) \vee \bigvee_{i=r+1}^{k} \alpha_i(a_1)$$

where $m \leq r$ and by our earlier assumption $0 < m, r < k$. Then the required inequality becomes

$$\frac{\int (\sum_{i=1}^{m} x_i)(\sum_{i=1}^{r} x_i)^{n+1} \, d\mu}{\int (\sum_{i=1}^{m} x_i)(\sum_{i=1}^{r} x_i)^{n} \, d\mu} \geq \frac{\int (\sum_{i=1}^{m} x_i + \sum_{i=r+1}^{k} x_i)(\sum_{i=1}^{r} x_i)^{n+1} \, d\mu}{\int (\sum_{i=1}^{m} x_i + \sum_{i=r+1}^{k} x_i)(\sum_{i=1}^{r} x_i)^{n} \, d\mu},$$

equivalently

$$\frac{\int (\sum_{i=1}^{m} x_i)(\sum_{i=1}^{r} x_i)^{n+1} \, d\mu}{\int (\sum_{i=1}^{m} x_i)(\sum_{i=1}^{r} x_i)^{n} \, d\mu} \geq \frac{\int (\sum_{i=r+1}^{k} x_i)(\sum_{i=1}^{r} x_i)^{n+1} \, d\mu}{\int (\sum_{i=r+1}^{k} x_i)(\sum_{i=1}^{r} x_i)^{n} \, d\mu}.$$

By Ax, for $1 \leq j \leq r$ and $s \in \mathbb{N}$,

$$m \int x_j \left(\sum_{i=1}^{r} x_i\right)^s d\mu = \int \left(\sum_{i=1}^{m} x_i\right) \left(\sum_{i=1}^{r} x_i\right)^s d\mu$$

and for $r + 1 \leq j \leq 2^q$

$$(k - r) \int x_j \left(\sum_{i=1}^{r} x_i\right)^s d\mu = \int \left(\sum_{i=r+1}^{k} x_i\right) \left(\sum_{i=1}^{r} x_i\right)^s d\mu.$$

Hence it suffices to show that

$$\frac{\int (\sum_{i=1}^{r} x_i)(\sum_{i=1}^{r} x_i)^{n+1} \, d\mu}{\int (\sum_{i=1}^{r} x_i)(\sum_{i=1}^{r} x_i)^{n} \, d\mu} \geq \frac{\int (\sum_{i=r+1}^{2^q} x_i)(\sum_{i=1}^{r} x_i)^{n+1} \, d\mu}{\int (\sum_{i=r+1}^{2^q} x_i)(\sum_{i=1}^{r} x_i)^{n} \, d\mu},$$

equivalently

$$\frac{\int (\sum_{i=1}^{r} x_i)^{n+2} \, d\mu}{\int (\sum_{i=1}^{r} x_i)^{n+1} \, d\mu} \geq \frac{\int (\sum_{i=1}^{r} x_i)^{n+1} \, d\mu - \int (\sum_{i=1}^{r} x_i)^{n+2} \, d\mu}{\int (\sum_{i=1}^{r} x_i)^{n} \, d\mu - \int (\sum_{i=1}^{r} x_i)^{n+1} \, d\mu}.$$

This amounts to

$$\frac{\int (\sum_{i=1}^{r} x_i)^{n+2} \, d\mu}{\int (\sum_{i=1}^{r} x_i)^{n+1} \, d\mu} \geq \frac{\int (\sum_{i=1}^{r} x_i)^{n+1} \, d\mu}{\int (\sum_{i=1}^{r} x_i)^{n} \, d\mu}. \tag{11}$$

which holds by EPIR.

Finally if equality held in the theorem for some n then we would have equality in (11) for some n, equivalently

$$\int \left((\sum_{i=1}^{r} x_i) - \frac{\int (\sum_{i=1}^{r} x_i)^{n+1} d\mu}{\int (\sum_{i=1}^{r} x_i)^{n} d\mu} \right)^2 (\sum_{i=1}^{r} x_i)^n d\mu = 0. \quad (12)$$

Let

$$a = \frac{\int (\sum_{i=1}^{r} x_i)^{n+1} d\mu}{\int (\sum_{i=1}^{r} x_i)^{n} d\mu}.$$

First consider $n = 0$. For (12) to hold, $\sum_{i=1}^{r} x_i$ must be equal to a for μ-almost all $\vec{x}$. Since μ is invariant under permutations of coordinates, the same must be true for the sum of any r of the x_i. Consequently, for μ-almost all $\vec{x}$, the sum of any r of the x_i must be a. Any two coordinates of such an $\vec{x}$ must be equal so μ is the discrete measure giving all the weight to $\vec{x} = \langle 2^{-q}, 2^{-q}, \ldots, 2^{-q} \rangle$ and hence $w = c_\infty^{L_q}$.

When $n \neq 0$, for (12) to hold $\sum_{i=1}^{r} x_i$ must be equal to a or to 0 for μ-almost all $\vec{x}$. Again, since μ is invariant under permutations of coordinates, the same must be true for the sum of any r of the x_i. Consequently, for μ-almost all $\vec{x}$, the sum of any r of the x_i must be a or 0. Any two coordinates of such an $\vec{x}$ must be either equal or differ by a. This is only possible when as before, $a = r2^{-q}$ and $\vec{x} = \langle 2^{-q}, 2^{-q}, \ldots, 2^{-q} \rangle$, or when $a = 1$ and $\vec{x} = \langle 0, 0, \ldots, 0, 1, 0, \ldots, 0, 0 \rangle$. Since w satisfies Ax, it follows that $w = c_\infty^{L}$ or $w = c_0^{L_q}$. □

We remark that the theorem does imply that with the exception of c_∞^{L} and $c_0^{L_q}$, any w satisfying Ax+Ex also satisfies all of (8), (9) and (10): we take $S(x)$ for $\phi(x)$ in both cases and $S(x) \to F(x)$, $S(x) \leftrightarrow F(x)$ or $F(x) \wedge S(x)$ respectively for $\theta(x)$ and note that $\theta(x) \wedge \phi(x)$ is logically equivalent to $F(x) \wedge S(x)$.

Again any argument against the evidence in (8), (9) and (10) providing a justification for $F(h)$ would seem to require one to hold the view that $c_\infty^{L_2}$ or $c_0^{L_2}$ was an acceptable choice. With $c_\infty^{L_2}$ there is the same counter as there was with Theorem 2 and using (4) a similar one can clearly be formulated for $c_0^{L_2}$.

Assuming Ex+Px+SN

Despite Johnson and Carnap's acceptance of Ax it seems that this is quite a step beyond Ex+Px+SN as far as being self evidently rational. For this reason it would be good if (8), (9) and (10) followed from just Ex+Px+SN since it would strengthen any claim as to their 'rationality'.

Treating (8) first:

Theorem 2. *Let w satisfy Ex+Px+SN. Then (8) holds for $n = 0, 1$, indeed*

$$w(F(h) \mid \bigwedge_{i=1}^{2}(S(k_i) \to F(k_i)) \wedge S(h)) \geq w(F(h) \mid (S(k_1) \to F(k_1)) \wedge S(h)) \quad (13)$$

$$\geq w(F(h) \mid S(h)) = 1/2 \quad (14)$$

with equality in (13) just if $w = c_\infty^{L^2}$ or $w = c_0^{L^2}$ and equality in (14) just if $w = c_\infty^{L^2}$.

Proof. The second inequality above is (5), and the result was proved in [11].

Turning to (13), $\alpha_1(x) = S(x) \wedge F(x)$, $\alpha_2(x) = S(x) \wedge \neg F(x)$, $\alpha_3(x) = \neg S(x) \wedge F(x)$, $\alpha_4(x) = \neg S(x) \wedge \neg F(x)$. We need to show that

$$\frac{w(\alpha_1(h) \wedge (\alpha_1(k_1) \vee \alpha_3(k_1) \vee \alpha_4(k_1)) \wedge (\alpha_1(k_2) \vee \alpha_3(k_2) \vee \alpha_4(k_2)))}{w((\alpha_1(h) \vee \alpha_2(h)) \wedge (\alpha_1(k_1) \vee \alpha_3(k_1) \vee \alpha_4(k_1)) \wedge (\alpha_1(k_2) \vee \alpha_3(k_2) \vee \alpha_4(k_2)))}$$

is greater or equal to

$$\frac{w(\alpha_1(h) \wedge (\alpha_1(k_1) \vee \alpha_3(k_1) \vee \alpha_4(k_1)))}{w((\alpha_1(h) \vee \alpha_2(h)) \wedge (\alpha_1(k_1) \vee \alpha_3(k_1) \vee \alpha_4(k_1)))}.$$

We can record this economically as

$$\frac{w(\alpha_1(\alpha_1 + \alpha_3 + \alpha_4)^2)}{w((\alpha_1 + \alpha_2)(\alpha_1 + \alpha_3 + \alpha_4)^2)} \geq \frac{w(\alpha_1(\alpha_1 + \alpha_3 + \alpha_4))}{w((\alpha_1 + \alpha_2)(\alpha_1 + \alpha_3 + \alpha_4))};$$

an equivalent inequality is

$$\frac{w(\alpha_1(\alpha_1 + \alpha_3 + \alpha_4)^2)}{w(\alpha_2(\alpha_1 + \alpha_3 + \alpha_4)^2)} \geq \frac{w(\alpha_1(\alpha_1 + \alpha_3 + \alpha_4))}{w(\alpha_2(\alpha_1 + \alpha_3 + \alpha_4))}.$$

Noting that by Ex, for distinct constants $a_{i_1}, a_{i_2}, a_{i_3}$ and $s, v, r \in \{1, 2, 3, 4\}$

$$w(\alpha_s(a_{i_1}) \wedge \alpha_v(a_{i_2}) \wedge \alpha_r(a_{i_3}))$$

depends only on s, v, r, we record this value as $w(\alpha_s \alpha_v \alpha_r)$ or, when for example $s = r$ as $w(\alpha_s^2 \alpha_v)$ etc. Let $p = w(\alpha_1^3)$, $y = w(\alpha_1^2 \alpha_2)$, $t = w(\alpha_1^2 \alpha_4)$, $z = w(\alpha_1 \alpha_2 \alpha_3)$. By Px+SN we have furthermore

$$p = w(\alpha_2^3) = w(\alpha_3^3) = w(\alpha_4^3),$$

$$y = w(\alpha_1^2 \alpha_3) = w(\alpha_4^2 \alpha_2) = w(\alpha_4^2 \alpha_3) = w(\alpha_2^2 \alpha_1) = w(\alpha_2^2 \alpha_4) = w(\alpha_3^2 \alpha_1) = w(\alpha_3^2 \alpha_4),$$

$$t = w(\alpha_4^2 \alpha_1) = w(\alpha_2^2 \alpha_3) = w(\alpha_3^2 \alpha_2),$$

$$z = w(\alpha_1 \alpha_2 \alpha_4) = w(\alpha_1 \alpha_3 \alpha_4) = w(\alpha_2 \alpha_3 \alpha_4)$$

so any $w(\alpha_i \alpha_j \alpha_k)$ is given by one of p, y, t, z.

Consequently for example

$$w(\alpha_1 \alpha_2) = w(\alpha_1 \alpha_2 (\alpha_1 + \alpha_2 + \alpha_3 + \alpha_4)) = 2y + 2z,$$

and

$$w(\alpha_1 \alpha_4) = 2t + 2z, \quad w(\alpha_1^2) = p + t + 2y,$$

so since we also have $w(\alpha_1 \alpha_3) = w(\alpha_1 \alpha_2) = w(\alpha_2 \alpha_4)$ and $w(\alpha_1 \alpha_4) = w(\alpha_2 \alpha_3)$ the required inequality becomes

$$\frac{p + 3t + 3y + 2z}{t + 2y + 6z} \geq \frac{p + 3t + 4y + 4z}{2t + 4y + 6z},$$

which simplifies to

$$pt + 2py + 4y^2 + 3t^2 + 8yt \geq 12z^2 + 6yz. \tag{15}$$

We now make two claims which will be proved later:

<u>Claim 1:</u> $p \geq y$ with equality just when $w = c_\infty^{L_2}$.

<u>Claim 2:</u> $y + t \geq 2z$.

Returning to the inequality (15), from Claim 2 we have $3(y+t)^2 \geq 12z^2$ and $y(y+t) \geq 2yz$ so

$$pt + 2py + 4y^2 + 3t^2 + 8yt \geq pt + 2py + 12z^2 + 2yz + yt.$$

But by Claims 1 and 2 we also have

$$pt + 2py + yt \geq yt + 2y^2 + yt = 2y(y+t) \geq 4yz, \tag{16}$$

so (15) follows.

It remains to show the Claims 1 and 2 and to consider when equality occurs in (15). Fortunately as the Claims are purely linear we can use the representation theorem which tells us that any probability function on the language $\{S, F\}$ satisfying Ex+Px+SN can be expressed as an integral (see [6, Lemma 6], dropping the redundant AP) using the probability functions

$$8^{-1}(w_{\langle x_1, x_2, x_3, x_4 \rangle} + w_{\langle x_1, x_3, x_2, x_4 \rangle} + w_{\langle x_4, x_2, x_3, x_1 \rangle} + w_{\langle x_4, x_3, x_2, x_1 \rangle}$$

$$+ w_{\langle x_2,x_1,x_4,x_3\rangle} + w_{\langle x_2,x_4,x_1,x_3\rangle} + w_{\langle x_3,x_1,x_4,x_2\rangle} + w_{\langle x_3,x_4,x_1,x_2\rangle}), \tag{17}$$

where the x_i are nonnegative real numbers summing to 1 and

$$w_{\langle x_i,x_j,x_k,x_r\rangle}(\alpha_1^{m_1}\alpha_2^{m_2}\alpha_3^{m_3}\alpha_4^{m_4}) = x_i^{m_1} x_j^{m_2} x_k^{m_3} x_r^{m_4}$$

Hence it is enough to show that $p \geq y$ and $y + t \geq 2z$ hold for this probability function. But these amount, respectively, to

$$2(x_1^3 + x_2^3 + x_3^3 + x_4^3) \geq x_1^2 x_2 + x_1^2 x_3 + x_2^2 x_1 + x_2^2 x_4 + x_3^2 x_1 + x_3^2 x_4 + x_4^2 x_2 + x_4^2 x_3,$$

$\iff$

$$\left.\begin{array}{l}(x_1 - x_2)^2(x_1 + x_2) + (x_1 - x_3)^2(x_1 + x_3) \\ + (x_4 - x_2)^2(x_4 + x_2) + (x_4 - x_3)^2(x_4 + x_3)\end{array}\right\} \geq 0,$$

with equality just when $x_1 = x_2 = x_3 = x_4$, and

$$x_1^2(x_2+x_3+2x_4) + x_2^2(x_1+x_4+2x_3) + x_3^2(x_1+x_4+2x_2) + x_4^2(x_2+x_3+2x_1)$$
$$\geq 2(2x_1x_2x_3 + 2x_2x_1x_4 + 2x_3x_1x_4 + 2x_4x_2x_3)$$

$\iff$

$$\left.\begin{array}{l}x_2(x_1-x_3)^2 + x_2(x_4-x_3)^2 + x_3(x_1-x_2)^2 + x_3(x_4-x_2)^2 \\ + x_1(x_2-x_4)^2 + x_1(x_3-x_4)^2 + x_4(x_2-x_1)^2 + x_4(x_3-x_1)^2\end{array}\right\} \geq 0.$$

with equality just when $x_1 = x_2 = x_3 = x_4$ or one of the x_i is 1.

Hence both claims hold and equality in Claim 1 can occur only when $w = c_\infty^{L_2}$ whilst in Claim 2 equality occurs just when the mixing measure featuring in the above mentioned representation of w gives measure 1 to the set of functions (17) with $x_1 = x_2 = x_3 = x_4$ or with one of the x_i equal to 1.

It follows that the first inequality in (16) is strict unless $t = y = 0$ or $w = c_\infty^{L_2}$. The former happens just when $w = c_0^{L_2}$ and hence equality in (15) occurs just when w is one of Carnap's $c_0^{L_2}$ and $c_\infty^{L_2}$.[14]

$\square$

---

[14] see e.g. [12].

Contrary to expectations however (8) can fail under Ex+Px+SN for $n > 2$. A counter-example is provided by a function of the form (17), with suitable x_1, x_2, x_3, x_4. For given x_1, x_2, x_3, x_4 (to be specified later) let $\tilde{w}$ be

$$8^{-1}(w_{\langle x_1,x_2,x_3,x_4 \rangle} + w_{\langle x_1,x_3,x_2,x_4 \rangle} + w_{\langle x_4,x_2,x_3,x_1 \rangle} + w_{\langle x_4,x_3,x_2,x_1 \rangle}$$
$$+ w_{\langle x_2,x_1,x_4,x_3 \rangle} + w_{\langle x_2,x_4,x_1,x_3 \rangle} + w_{\langle x_3,x_1,x_4,x_2 \rangle} + w_{\langle x_3,x_4,x_1,x_2 \rangle})$$

and define

$$R(n) = \frac{\tilde{w}(\alpha_1(\alpha_1 + \alpha_3 + \alpha_4)^n)}{(\tilde{w}(\alpha_1 + \alpha_2)(\alpha_1 + \alpha_3 + \alpha_4)^n)}$$
$$= \tilde{w}\left(F(h) \mid \bigwedge_{i=1}^{n}(S(k_i) \to F(k_i)) \wedge S(h)\right). \qquad (18)$$

Write $A = x_1 + x_4$ and $B = x_2 + x_3$. We have

$$w_{\langle x_1,x_2,x_3,x_4 \rangle}(\alpha_1(\alpha_1 + \alpha_3 + \alpha_4)^n) = x_1(A + x_3)^n,$$
$$w_{\langle x_1,x_2,x_3,x_4 \rangle}((\alpha_1 + \alpha_2)(\alpha_1 + \alpha_3 + \alpha_4)^n) = (x_1 + x_2)(A + x_3)^n$$

etc. so collecting terms, $R(n)$ is

$$\frac{A(A+x_3)^n + A(A+x_2)^n + B(B+x_1)^n + B(B+x_4)^n}{(A+2x_2)(A+x_3)^n + (A+2x_3)(A+x_2)^n + (B+2x_4)(B+x_1)^n + (B+2x_1)(B+x_4)^n}.$$

Multiplying $R(n+1) \geq R(n)$ by the denominators and subtracting the RHS from the LHS shows it equivalent to $C - D \geq 0$ where C is the sum of products of terms from these two columns:

$$\begin{array}{ll}
A(A+x_3)^{n+1} & (A+2x_2)(A+x_3)^n \\
A(A+x_2)^{n+1} & (A+2x_3)(A+x_2)^n \\
B(B+x_1)^{n+1} & (B+2x_4)(B+x_1)^n \\
B(B+x_4)^{n+1} & (B+2x_1)(B+x_4)^n
\end{array}$$

and D is the sum of products in these two columns:

$$\begin{array}{ll}
A(A+x_3)^n & (A+2x_2)(A+x_3)^{n+1} \\
A(A+x_2)^n & (A+2x_3)(A+x_2)^{n+1} \\
B(B+x_1)^n & (B+2x_4)(B+x_1)^{n+1} \\
B(B+x_4)^n & (B+2x_1)(B+x_4)^{n+1}
\end{array}$$

This amounts to the sum of the following 12 terms being non-negative:

$$(A+x_3)^n \ (A+x_2)^n \ A \ (A+2x_3) \ (x_3-x_2)$$
$$(A+x_3)^n \ (B+x_4)^n \ A \ (B+2x_1) \ (A+x_3-B-x_4)$$
$$(A+x_3)^n \ (B+x_1)^n \ A \ (B+2x_4) \ (A+x_3-B-x_1)$$
$$(A+x_2)^n \ (A+x_3)^n \ A \ (A+2x_2) \ (x_2-x_3)$$
$$(A+x_2)^n \ (B+x_1)^n \ A \ (B+2x_4) \ (A+x_2-B-x_1)$$
$$(A+x_2)^n \ (B+x_4)^n \ A \ (B+2x_1) \ (A+x_2-B-x_4)$$
$$(B+x_4)^n \ (A+x_3)^n \ B \ (A+2x_2) \ (B+x_4-A-x_3)$$
$$(B+x_4)^n \ (A+x_2)^n \ B \ (A+2x_3) \ (B+x_4-A-x_2)$$
$$(B+x_4)^n \ (B+x_1)^n \ B \ (B+2x_4) \ (x_4-x_1)$$
$$(B+x_1)^n \ (A+x_3)^n \ B \ (A+2x_2) \ (B+x_1-A-x_3)$$
$$(B+x_1)^n \ (A+x_2)^n \ B \ (A+2x_3) \ (B+x_1-A-x_2)$$
$$(B+x_1)^n \ (B+x_4)^n \ B \ (B+2x_1) \ (x_1-x_4)$$

Combining the obvious pairs yields the sum of the following six terms

$$(A+x_3)^n \ (A+x_2)^n \ A \ 2(x_3-x_2)^2$$
$$(A+x_3)^n \ (B+x_4)^n \ 2(Ax_1-Bx_2) \ (A+x_3-B-x_4)$$
$$(A+x_3)^n \ (B+x_1)^n \ 2(Ax_4-Bx_2) \ (A+x_3-B-x_1)$$
$$(A+x_2)^n \ (B+x_1)^n \ 2(Ax_4-Bx_3) \ (A+x_2-B-x_1)$$
$$(A+x_2)^n \ (B+x_4)^n \ 2(Ax_1-Bx_3) \ (A+x_2-B-x_4)$$
$$(B+x_4)^n \ (B+x_1)^n \ B \ 2(x_4-x_1)^2$$

Rewriting A and B back in terms of the x_i, this is the sum of

$$(1-x_2)^n \ (1-x_3)^n \ (x_1+x_4) \ 2(x_3-x_2)^2$$
$$(1-x_2)^n \ (1-x_1)^n \ 2(x_1^2-x_2^2+x_1x_4-x_2x_3) \ (x_1-x_2)$$
$$(1-x_2)^n \ (1-x_4)^n \ 2(x_4^2-x_2^2+x_1x_4-x_2x_3) \ (x_4-x_2)$$
$$(1-x_3)^n \ (1-x_4)^n \ 2(x_4^2-x_3^2+x_1x_4-x_2x_3) \ (x_4-x_3)$$
$$(1-x_3)^n \ (1-x_1)^n \ 2(x_1^2-x_3^2+x_1x_4-x_2x_3) \ (x_1-x_3)$$
$$(1-x_1)^n \ (1-x_4)^n \ (x_2+x_3) \ 2(x_4-x_1)^2$$

Let $\epsilon > 0$ and

$$x_1 = 1-6\epsilon, \quad x_2 = 3\epsilon, \quad x_3 = 2\epsilon, \quad x_4 = \epsilon.$$

Then $1-x_1$ and x_2, x_3, x_4 are of order ϵ, so the second, fifth and sixth products are of order ϵ^n whilst the first, third and fourth are respectively

$$(1-3\epsilon)^n \ (1-2\epsilon)^n \ (1-5\epsilon) \ 2\epsilon^2$$

$$(1-3\epsilon)^n \ (1-\epsilon)^n \ 2(\epsilon-20\epsilon^2) \ (-2\epsilon)$$

$$(1-2\epsilon)^n \ (1-\epsilon)^n \ 2(\epsilon - 15\epsilon^2) \ (-\epsilon)$$

so their sum is negative of order ϵ^2. Hence for $n > 2$ and a sufficiently small ϵ this is a counterexample to $R(n+1) \geq R(n)$.

We remark that the very same approach as above does not work with $n = 2$ (3 kitchens) because with these $x_1, \ldots, x_4$ all but the last product are of the order ϵ^2 (the last one being of order ϵ^3) and their sum for small ϵ is positive.

According to the natural requirement then that more examples should provide an ever greater enhancement of the probability of $F(s)$ this argues against the implication formalisation, at least if one wants to limit the assumptions to Ex+Px+SN rather than the stronger Ex+Ax. (As we shall see shortly this same criticism applies to the conjunction formalization, but not to the bi-implication formalisation.)

It is worth noting at this point that for probability functions of the form (17), if x_4 is the strictly smallest of the x_i then

$$\lim_{n \to \infty} w(F(h) \,|\, \bigwedge_{i=1}^{n}(S(k_i) \to F(k_i)) \wedge S(h)) = \frac{x_2 + x_3}{x_2 + x_3 + 2x_4}, \qquad (19)$$

so not 1 but greater than $1/2$.

The status of the one remaining case, when $n = 2$, the '3 kitchens problem', is open. Given these counter-examples one might wonder if it was possible that Ex+Px+SN was not enough to even justify jumping to the conclusion $F(h)$ on the basis of more than 2 kitchen fires. Fortunately that recommendation is still good:

Theorem 3. *For w satisfying Ex+Px+SN, and any $n \geq 1$*

$$w(F(a_{n+1}) \,|\, S(a_{n+1}) \wedge \bigwedge_{i=1}^{n}(S(a_i) \to F(a_i))) \geq 1/2 = w(F(a_{n+1}) \,|\, S(a_{n+1})),$$

with equality just if $w = c_\infty^{L_2}$.

Proof. Using the usual de Finetti Representation Theorem for w a probability function on L_2 satisfying Ex, the required inequality becomes

$$\frac{\int x_1(x_1 + x_3 + x_4)^n \, d\mu(\vec{x})}{\int (x_1 + x_2)(x_1 + x_3 + x_4)^n \, d\mu(\vec{x})} \geq \frac{1}{2}.$$

Simplifying gives

$$\int x_1(x_1 + x_3 + x_4)^n \, d\mu(\vec{x}) \geq \int x_2(x_1 + x_3 + x_4)^n \, d\mu(\vec{x}),$$

equivalently
$$\int (x_1 - x_2)(1 - x_2)^n \, d\mu(\vec{x}) \geq 0.$$

By the trick in [12, page 90] we can assume that the measure μ is invariant under those permutations of coordinates which are 'licensed' by SN+Px, in particular the permutation transposing x_1, x_2 (and x_3, x_4). Hence

$$\int (x_1 - x_2)(1 - x_2)^n \, d\mu(\vec{x}) = \int (x_2 - x_1)(1 - x_1)^n \, d\mu(\vec{x})$$

and it is enough to show that

$$\int (x_1 - x_2)(1 - x_2)^n + (x_2 - x_1)(1 - x_1)^n \, d\mu(\vec{x}) \geq 0.$$

But the polynomial being integrated here is just

$$(x_1 - x_2)((1 - x_2)^n - (1 - x_1)^n)$$

which equals

$$(x_1 - x_2)^2 \sum_{i=0}^{n-1} (1 - x_1)^{n-1-i} (1 - x_2)^i \geq 0 \qquad (20)$$

so the result clearly holds.

Finally we can only have equality in (20) for all support points of μ if $x_1 = x_2$ (and perforce $x_1 = x_3 = x_4$ by the assumed invariance of μ under permutations licensed by Px+SN) for all support points so the only possible support point is $\langle 4^{-1}, 4^{-1}, 4^{-1}, 4^{-1} \rangle$ and w on this sublanguage must be $c_\infty^{L_2}$.

$\square$

On a more positive note however we can fully answer this question when it comes to (9):

Theorem 4. *For w a probability function on L_2 satisfying Ex+Px+SN and $n \geq 0$,*

$$w(F(h) \mid \bigwedge_{i=1}^{n+1} (S(k_i) \leftrightarrow F(k_i)) \wedge S(h)) \geq w(F(h) \mid \bigwedge_{i=1}^{n} (S(k_i) \leftrightarrow F(k_i)) \wedge S(h)). \quad (21)$$

We remark that equality does hold for some special probability functions but they can be dismissed on similar grounds as $c_\infty^{L_2}$ and $c_0^{L_2}$, see below.

Proof. Using the same notation as above, we need to show that for a probability function w satisfying Ex+Px+SN and $n \geq 0$,

$$\frac{w(\alpha_1(\alpha_1 + \alpha_4)^{n+1})}{w((\alpha_1 + \alpha_2)(\alpha_1 + \alpha_4)^{n+1})} \geq \frac{w(\alpha_1(\alpha_1 + \alpha_4)^n)}{w((\alpha_1 + \alpha_2)(\alpha_1 + \alpha_4)^n)}, \tag{22}$$

equivalently that

$$\frac{w(\alpha_1(\alpha_1 + \alpha_4)^n)}{w(\alpha_2(\alpha_1 + \alpha_4)^n)}$$

is non-decreasing. To this end, it suffices to show that

$$\frac{w(\alpha_1(\alpha_1 + \alpha_4)^n)}{w((\alpha_1 + \alpha_4)^n)}$$

is non decreasing and

$$\frac{w(\alpha_2(\alpha_1 + \alpha_4)^n)}{w((\alpha_1 + \alpha_4)^n)}$$

is non-increasing. This follows by EPIR since by SN+Px

$$w(\alpha_1(\alpha_1 + \alpha_4)^n) = 2^{-1} w((\alpha_1 + \alpha_4)(\alpha_1 + \alpha_4)^n)$$

and

$$\begin{aligned} w(\alpha_2(\alpha_1 + \alpha_4)^n) &= 2^{-1} w((\alpha_2 + \alpha_3)(\alpha_1 + \alpha_4)^n) \\ &= 2^{-1}(w((\alpha_1 + \alpha_4)^n) - w((\alpha_1 + \alpha_4)(\alpha_1 + \alpha_4)^n)). \end{aligned}$$

□

As in the corresponding proof[15] of Theorem 1 if we have equality in (22) for some n and $n+1$ then either every point in the support of μ must be of one of the form $\langle x_1, x_2, (1/2) - x_2, (1/2) - x_1 \rangle$ or every point in the support of μ must be of the form $\langle 0, x_2, 1 - x_2, 0 \rangle$ or $\langle x_1, 0, 0, 1 - x_1 \rangle$ and $n > 0$.

These conditions are not enough to force w to be one of $c_\infty^{L_2}$ or $c_0^{L_2}$. However a similar argument can be made to rebuke the probability functions w which do give equality here. Namely they would have to satisfy

$$w((\alpha_1 + \alpha_4) \,|\, (\alpha_1 + \alpha_4)^n) = w((\alpha_1 + \alpha_4) \,|\, (\alpha_1 + \alpha_4))$$

for all $n > 0$.

---
[15] Arguing about the probability function for the language with one predicate R which we obtain from w upon replacing $\alpha_1 \vee \alpha_4$ by R and $\alpha_2 \vee \alpha_3$ by $\neg R$.

Turning now to the formalisation as a conjunction, as we shall see shortly from Theorem 5 we do have that (10) holds for $n = 0$ when w satisfies Ex+Px+SN. However that is not generally the case for $n \geq 2$: The probability function $(1 - \epsilon)\omega_1 + \epsilon\omega_2$ satisfies Ex+Px+SN but not (10) when $\epsilon > 0$ is small, ω_1 is of the form (17) with $x_1 = x_4 = 1/2$, $x_2 = 0 = x_3$ and ω_2 is also of this form but with $x_1 = 3/4$, $x_2 = 1/4$, $x_3 = x_4 = 0$. As with implication (but not bi-implication) these counter-examples furnish a criticism of the conjunction formalisation.

To our knowledge the current status of (10) for $n = 1$ is an open problem.

Fortunately we do, as with implication, have:

Theorem 5. *For w a probability function on L_2 satisfying Ex+Px+SN, and $n \geq 1$*

$$w(F(a_{n+1}) \,|\, S(a_{n+1}) \wedge \bigwedge_{i=1}^{n}(S(a_i) \wedge F(a_i))) \geq 1/2 = w(F(a_{n+1}) \,|\, S(a_{n+1})),$$

with equality just if $w = c_\infty^{L_2}$.

Proof. Proceeding as in the proof of Theorem 3, we use the usual de Finetti Representation Theorem for w a probability function on L_2 satisfying Ex. The required inequality becomes

$$\frac{\int x_1^{n+1}\, d\mu(\vec{x})}{\int (x_1 + x_2) x_1^n\, d\mu(\vec{x})} \geq \frac{1}{2},$$

equivalently

$$\int x_1^{n+1}\, d\mu(\vec{x}) \geq \int x_2 x_1^n\, d\mu(\vec{x}),$$

that is,

$$\int (x_1 - x_2)\, x_1^n\, d\mu(\vec{x}) \geq 0.$$

Again since w satisfies also Px we can assume that the measure μ is invariant under the permutation transposing x_1, x_2 and x_3, x_4. Hence

$$\int (x_1 - x_2)\, x_1^n\, d\mu(\vec{x}) = \int (x_2 - x_1)\, x_2^n\, d\mu(\vec{x})$$

and it is enough to show that

$$\int \left((x_1 - x_2)\, x_1^n + (x_2 - x_1)\, x_2^n \right) d\mu(\vec{x}) \geq 0.$$

But the polynomial being integrated here is just

$$(x_1 - x_2)(x_1^n - x_2^n)$$

which is clearly nonnegative, so the result follows. The last part about $c_\infty^{L_2}$ follows as in Theorem 3. □

Mill's Property

In [9, Vol.7,p186] the Scotish philosopher J.S.Mill suggested (as others have since) that when we use for example

All men are mortal

to conclude that the Duke of Wellington is mortal it is not that we already know all instances of this universal but that we know a sufficient number of them to feel justified in saving mental storage space by rounding up our knowledge to 'All men are mortal'. In other words we are transforming an argument by induction into a fully deductive argument. From this viewpoint then the reality of the Indian Schema for one reasoning to oneself might be read as:

(a) *In the many cases I have experienced of smoke there has invariably been fire.*

(b) *There is smoke on the hill.*

(c) *Therefore there is fire on the hill.*

If such reasoning can be taken to be in some sense 'rational' then it suggests we should investigate the status within PIL of probability functions w on L_q satisfying the somewhat more general principle:

Mill's Property, MP

For $\theta(a_1), \phi(a_1) \in QFSL_q$ with $w(\theta(a_1) \wedge \phi(a_1)) > 0$,

$$lim_{n \to \infty} w \left(\theta(a_{n+1}) \,|\, \phi(a_{n+1}) \wedge \bigwedge_{i=1}^{n} \theta(a_i) \right) = 1$$

Theorem 6. *Let w be a probability function on L_q satisfying Ax and with de Finetti prior μ. Then w satisfies MP just if all the points $\langle 0, 0, \ldots, 0, 1, 0, \ldots, 0, 0 \rangle$ are in the support of μ.*

Proof. First suppose that $\vec{x} = \langle 1, 0, 0, \ldots, 0 \rangle$ is not in the support of μ, say that $\mu(A_\delta) = 0$ where $\delta > 0$ and

$$A_\delta = \{ \vec{y} \in \mathbb{D}_{2^q} \,|\, |\vec{y} - \vec{x}| < \delta \}.$$

Then

$$w(\alpha_1(a_{n+1}) \mid \bigwedge_{i=1}^{n} \alpha_1(a_i)) = \frac{\int_{\mathbb{D}_{2^q}} x_1^{n+1} \, d\mu}{\int_{\mathbb{D}_{2^q}} x_1^n \, d\mu}$$

$$= \frac{\int_{\mathbb{D}_{2^q} - A_\delta} x_1^{n+1} \, d\mu}{\int_{\mathbb{D}_{2^q} - A_\delta} x_1^n \, d\mu}$$

$$\leq \frac{\int_{\mathbb{D}_{2^q} - A_\delta} (1-\delta) x_1^n \, d\mu}{\int_{\mathbb{D}_{2^q} - A_\delta} x_1^n \, d\mu} < 1,$$

so MP fails for this θ and $\phi = \top$, where $\top$ stands for a tautology.

In the other direction suppose that each of these points $\langle 0, 0, \ldots, 0, 1, 0, \ldots, 0, 0 \rangle$ is in the support of μ and let

$$\theta(a_1) \equiv \bigvee_{i=1}^{r} \alpha_i(a_1), \quad \phi(a_1) \equiv \bigvee_{i=1}^{m} \alpha_i(a_1) \vee \bigvee_{i=r+1}^{k} \alpha_i(a_1)$$

where $0 < m \leq r$. We may assume that $k \geq r+1$ otherwise the result is immediate.

We need to show that

$$\frac{\int (\sum_{i=r+1}^{k} x_i)(\sum_{i=1}^{r} x_i)^n \, d\mu}{\int (\sum_{i=1}^{m} x_i)(\sum_{i=1}^{r} x_i)^n \, d\mu}.$$

tends to zero as $n \to \infty$. Using Ax as in the proof of Theorem 1 it is enough to show that

$$\frac{\int (\sum_{i=r+1}^{2^q} x_i)(\sum_{i=1}^{r} x_i)^n \, d\mu}{\int (\sum_{i=1}^{r} x_i)^{n+1} \, d\mu} \quad (23)$$

tends to zero as $n \to \infty$.

Let $0 < \delta < \nu$ and

$$B_\delta = \{\vec{x} \in \mathbb{D}_{2^q} \mid \sum_{i=1}^{r} x_i \geq 1 - \delta\}$$

and similarly for ν. By the assumption of MP in the theorem $\mu(B_\delta) > 0$.

We can write (23) as

$$\frac{\int_{B_\nu} (\sum_{i=r+1}^{2^q} x_i)(\sum_{i=1}^{r} x_i)^n \, d\mu + \int_{\mathbb{D}_{2^q} - B_\nu} (\sum_{i=r+1}^{2^q} x_i)(\sum_{i=1}^{r} x_i)^n \, d\mu}{\int_{B_\delta} (\sum_{i=1}^{r} x_i)^{n+1} \, d\mu + \int_{B_\nu - B_\delta} (\sum_{i=1}^{r} x_i)^{n+1} \, d\mu + \int_{\mathbb{D}_{2^q} - B_\nu} (\sum_{i=1}^{r} x_i)^{n+1} \, d\mu}. \quad (24)$$

Since

$$\frac{\int_{B_\nu}(\sum_{i=r+1}^{2^q} x_i)(\sum_{i=1}^r x_i)^n \, d\mu}{\int_{B_\delta}(\sum_{i=1}^r x_i)^{n+1}\, d\mu + \int_{B_\nu - B_\delta}(\sum_{i=1}^r x_i)^{n+1}\, d\mu} \leq \frac{\nu}{1-\nu},$$

and

$$\frac{\int_{\mathbb{D}_{2^q} - B_\nu}(\sum_{i=r+1}^{2^q} x_i)(\sum_{i=1}^r x_i)^n \, d\mu}{\int_{B_\delta}(\sum_{i=1}^r x_i)^{n+1}\, d\mu} \leq \frac{(1-\nu)^n(1-\mu(B_\nu))}{(1-\delta)^{n+1}\mu(B_\delta)}$$

it follows that by choosing δ, ν sufficiently small and then n sufficiently large we can make (24) arbitrarily small, as required. $\square$

This proof has assumed Ax. If we only assume Ex+Px+SN then Mill's Property may not hold as is apparent from (19).

We remark that a corollary of Theorem 6 is that, assuming Ax and regularity (i.e. $w(\theta) > 0$ whenever $\theta \in QFSL_q$ is consistent), Reichenbach's Axiom, see [12] for a formulation in the notation of this paper, implies Mill's Property since by Theorem 15.1 of that monograph that axiom is equivalent to *every* point in $\mathbb{D}_{2^q}$ being a support point of μ.

The Lake

The version of the Indian Schema which we have considered here is based on Sūtra 36 which is commonly referred to as a 'homogeneous example'. We have variously formalised this as

$$S(k) \to F(k), \quad S(k) \leftrightarrow F(k) \tag{25}$$

or

$$S(k) \wedge F(k). \tag{26}$$

However in Sūtra 37[16] Gotama describes another sort of example, a heterogeneous example. According to S.C.Vidyabhusana's rendering of the Sūtra, [15, p12]:

> *A heterogeneous (or negative) example is a familiar instance which is known to be devoid of the property to be established and which implies that*

---

[16] *tad-viparyayād vā viparītam.*

the absence of this property is invariably rejected in the reason given.[17]

In our smoke-fire scenario a commonly stated such example is that of the lake [which is both fire and smoke free] which is combined with the homogeneous example of the kitchen to give the schema:

(a) *Where there is smoke there is fire, like in the kitchen and (un)like on the lake.*

(b) *There is smoke on the hill.*

(c) *Therefore there is fire on the hill.*

Exactly how we should formalise the lake (denoted l) example is not clear (to us) but it would seem that given the formalisations in (25) it could arguably be, respectively:

$$\neg F(l) \to \neg S(l), \quad \neg S(l) \leftrightarrow \neg F(l),$$

which are simply covered by the two kitchen version. For (26) it could arguably be

$$\neg F(l) \wedge \neg S(l).$$

However there are probability functions satisfying Ex+Ax for which the inequality

$$w(F(h) \,|\, S(h) \wedge S(k) \wedge F(k) \wedge \neg S(l) \wedge \neg F(l))$$

$$\geq w(F(h) \,|\, S(h) \wedge S(k) \wedge F(k)) \qquad (27)$$

does not hold.[18] On this evidence again then it seems that the appropriateness of capturing the example by a conjunction rather than an implication or bi-implication is questionable.

---

[17]Translations here from the original Sanskrit are considered notoriously difficult. We are grateful to Eberhard Guhe for suggesting the more literal

Or in the case opposite to that [i.e. the above-mentioned positive example] it [the udāharaṇa] is contrary [to the case at issue].

where the parenthetic comments have been added by him for clarification and are not part of the Sanskrit original.

[18]For w satisfying Ax (27) reduces to $2y^2 \geq xz + yz$ where $x = w(\alpha_1^3)$, $y = w(\alpha_1^2 \alpha_2)$ and $z = w(\alpha_1 \alpha_2 \alpha_3)$. This fails in the case of the probability function

$$4^{-1}(w_{\langle 1-3\epsilon, \epsilon, \epsilon, \epsilon \rangle} + w_{\langle \epsilon, 1-3\epsilon, \epsilon, \epsilon \rangle} + w_{\langle \epsilon, \epsilon, 1-3\epsilon, \epsilon \rangle} + w_{\langle \epsilon, \epsilon, \epsilon, 1-3\epsilon \rangle})$$

with $\epsilon > 0$ very small since it does satisfy Ax but gives, up to lowest powers of ϵ, $x = 1/4$, $y = \epsilon/4$, $z = 3\epsilon^2/4$ so $2y^2 < xz$.

Conclusion

In this paper we have limited ourselves to perhaps the most natural present day formulations of the Indian Schema, namely treating 'smoke', or 'smoky', as a predicate and 'hill' as a constant, etc. and have shown how most of these can claim to be justified as rational, at least if $c_0^{L_q}, c_\infty^{L_q}$ are excluded, on the basis of following from various symmetry principles in PIL. On this count taking the example to be a bi-implication seems to come out best if we assume only Ex+Px+SN but once we move up to Ex+Ax all three, implication, bi-implication and conjunction, fare satisfactorily.

Given the arcane complexities of Sanskrit however it is certainly not clear, even unlikely, that the formalisation presented here was how Gotama and the subsequent commentators on the Nyāyasūtra would have seen it. For example it has been suggested that 'hill' might have been thought of as a predicate and smoke, or smokiness, as a constant etc. or that they are both constants and the connection between them is via a binary relation A of 'happens at', see [8]. We plan to investigate these alternatives in a future paper but for the present we should emphasize that our primary purpose in this paper, and the earlier paper [11] which it extends, has not been to argue about what Gotama et al could have meant but rather to justify the rationality of the version of the schema as it seemed many Victorian (and later) readers dismissively understood it.

References

[1] Carnap, R., *The Continuum of Inductive Methods,* University of Chicago Press, 1952.

[2] Gaifman, H., Applications of de Finetti's Theorem to Inductive Logic, in *Studies in Inductive Logic and Probability,* Volume I, eds. R.Carnap & R.C.Jeffrey, University of California Press, 1971, pp235-251.

[3] Ganeri, J., ed., *Indian Logic: A Reader* Routledge, London & New York, 2001.

[4] Ganeri, J., Indian Logic and the Colonization of Reason, in *Indian Logic: A Reader,* ed. J.Ganeri, Routledge, London & New York, 2001, pp1-25.

[5] Ganeri, J., Ancient Indian Logic as a Theory of Case Based Reasoning, *Journal of Indian Philosophy,* **31**:33-45, 2003.

[6] Hill, A. & Paris, J.B., An Analogy Principle in Inductive Logic, *Annals of Pure and Applied Logic,* **164**, 2013, pp1293-1321.

[7] Johnson, W.E., Probability: The Deductive and Inductive Problems, *Mind,* 1932, **41**:409-423.

[8] Matilal, B.K., Introducing Indian Logic, in *Indian Logic: A Reader,* ed. J.Ganeri, Routledge, London & New York, 2001, pp183-215.

[9] Mill, J.S., A System of Logic, in *Collected Works*, vols. 7&8, University of Toronto Press, Toronto, 1973.

[10] Oetke, C., Ancient Indian Logic as a Theory of Non-monotonic Reasoning, *Journal of Indian Philosophy,* **24**:447-539, 1996.

[11] Paris, J.B. & Vencovská, A., The Indian Schema as Analogical Reasoning, submitted to the *Journal of Philosophical Logic.* Currently available at MIMS EPrints, http://eprints.ma.man.ac.uk/2438/

[12] Paris, J.B. & Vencovská, A., *Pure Inductive Logic*, in the Association of Symbolic Logic Perspectives in Mathematical Logic Series, Cambridge University Press, April 2015.

[13] Randle, H.N., A Note on the Indian Syllogism, *Mind*, 1924, **33**(132):398-414.

[14] *The Philosophy of Rudolf Carnap,* ed. P.A.Schlipp, Open Court, 1963.

[15] Vidyabhusana, S.C., *The Nyāya Sūtras of Gotama,* vol.8 in the *Sacred Books of the Hindus* series, Ed. B.D.Basu, published by the Panini Office, Bhuvaneswari Asrama, Bahadurgaj, Allababad, 1913. Republished by the AMS Press, INC., New York, 1974, ISBN:0-404-57808-X.

[16] Zilberman, D.B., *Analogy in Indian and Western Philosophical Thought*, Boston Studies in the Philosophy of Science vol.243, eds. H.Gourko & R.S.Cohen, 2006.

www.ingramcontent.com/pod-product-compliance
Lightning Source LLC
Chambersburg PA
CBHW081345040426
42450CB00015B/3312